ENVIRONMENTAL AND SOCIAL JUSTICE ISSUES

This book uses a declarative mapping methodology to examine a range of issues relating to environmental and social justice issues, including climate change, homelessness, refugees, food insecurity, and racial and gender inequality. The book explores how we can bring about change in order to have a meaningful impact on these problems, using a literature-based approach to identify and analyse this through the Declarative Mapping Method, showing how this methodology can be used in the context of these issues. The authors build a body of knowledge based upon published research, to offer a template that may be used to bring about meaningful and appropriate changes in human behaviour in a variety of social/ecological justice contexts. In a world where most of the global challenges we face are a result of human behaviour, the book applies psychological principles to gain a deeper understanding of our responses to world issues. Case studies are included to show how specific strategies can be used to address problems, and a holistic perspective offers strategies and insights into addressing these challenges. This is an ideal text for researchers and students interested in environmental and social issues, especially those looking to find ways to address them through research methodologies.

Dr. Peter Steidl is a business consultant, author, speaker, and educator. His main interest lies in understanding why we behave the way we do and finding ways to positively shape behaviour on an individual, organizational, and societal level.

Prof. Paul M. W. Hackett has wide experience in social science and humanities research and is the originator of the Declarative Mapping approach to qualitative research. He holds a PhD in psychology and a PhD in fine art. He has held appointments at several universities, including Oxford, Cambridge, Tufts, Harvard, Cardiff, and Durham.

Ava Gordley-Smith is a PhD research student at the University of Wales Trinity Saint David. Her research aims to develop a multidimensional psychological framework for understanding the relationship and intersection of attitudes toward environmental and social justice issues.

ENVIRONMENTAL AND SOCIAL JUSTICE ISSUES

A Declarative Mapping Literature-based Approach for Achieving Pro-ecological and Social Change

Peter Steidl, Paul M. W. Hackett and Ava Gordley-Smith

Routledge
Taylor & Francis Group

LONDON AND NEW YORK

Designed cover image: Getty © W.L. Davies

First published 2025
by Routledge
4 Park Square, Milton Park, Abingdon, Oxon OX14 4RN

and by Routledge
605 Third Avenue, New York, NY 10158

Routledge is an imprint of the Taylor & Francis Group, an informa business

British Library Cataloguing-in-Publication Data
A catalogue record for this book is available from the British Library

ISBN: 978-1-032-76112-1 (hbk)
ISBN: 978-1-032-76111-4 (pbk)
ISBN: 978-1-003-47716-7 (ebk)

DOI: 10.4324/9781003477167

Typeset in Optima
by SPi Technologies India Pvt Ltd (Straive)

CONTENTS

About the Authors   *vi*
Preface   *viii*

1 The Commitment Gap   1

2 We Can Succeed!   22

3 Human Nature: Ignore It at Your Peril!   38

4 Addressing Global Challenges   57

5 Progress Is a Puzzle: We Need to Get All the Pieces Right   93

6 Getting Ready to Act   109

7 Action! What Can You Do? What Can Others Do?   119

Index   *134*

ABOUT THE AUTHORS

In this initial section of our book we introduce ourselves, the authors. We do this to provide a context for readers to understand the writing that follows as well as to, in some ways, bracket our work and to position the content that follows.

Dr. Peter Steidl is a business consultant, author, speaker, and educator. His main interest lies in understanding why we behave the way we do and finding ways to positively shape behaviour on an individual, organizational, and societal level. He has been a consultant to Fortune Global 500 and other multinational corporations in more than 20 countries on five continents, and has held positions with the BBDO, WPP and Dentsu advertising networks. He has also served as a Temporary Advisor to the World Health Organization and has worked on mass behavioural change programmes on issues such as road safety, obesity, and cancer prevention. As well as being the author/co-author of 16 books, his articles, interviews, and commentaries have appeared in journals and trade publications and on television and radio in Europe, the Asia-Pacific, and the United States. He holds an MBA from the Vienna University of Economics and Business, and a PhD from the University of Vienna.

Prof. Paul M. W. Hackett has wide experience in social science and humanities research and is the originator of the Declarative Mapping approach to qualitative research. His research interests span many aspects of behaviour and experience, focussing on the investigation of ontologies and epistemologies that have originated in the African continent. He has around 275 publications, including more than 30 books. He is a visiting research professor in the department of philosophy at Nnamdi Azikiwe University, Nigeria, an honorary fellow in the philosophy department at University of Wales Trinity St David and a PhD supervisor at the same university, a visiting professor in health research methods at the University of Suffolk, and a professor in the school of communication at Emerson College. He holds a PhD in psychology and a PhD in fine art. He has held appointments at several universities, including Oxford, Cambridge, Tufts, Harvard, Cardiff and Durham.

Ava Gordley-Smith is a PhD research student at the University of Wales Trinity Saint David. Her research aims to develop a multidimensional psychological framework for understanding the relationship and intersection of attitudes towards environmental and social justice issues. Outside of her PhD research, Gordley-Smith focuses more broadly on both theoretical and applied methodologies in the social sciences, the creative disciplines, and communication studies. Gordley-Smith is also a partner and manager for mediaman, an international digital agency and the US country manager for the sustainable mobility venture studio Next Mobility Labs. In both of these roles, Gordley-Smith applies her academic research to direct strategic communications. Gordley-Smith received a Masters' degree from Emerson College's Strategic Marketing Communication programme, with her research focusing on utilizing ecology-centric and sociological methodologies as an approach to construct both marketing and brand strategies. Gordley-Smith is passionate about researching the intersectional inequalities and proposing change in gender, sex, sexuality, environmentalism, and race studies and is currently completing multiple book projects regarding these subjects.

PREFACE

We are tired of incompetence, self-interest, ignorance, arrogance, and other human traits that prevent us from addressing the major global crises we face. We are tired of self-important people talking about what should be done rather than doing it. And we are tired of people delegating their thinking to a political party, social media, or some moron who pretends that we don't have a problem – we just need to relax and enjoy the moment.

What upsets us most is that we know how to address our challenges. Indeed, we talk about this at workshops, conferences, meetings, in documentaries, media interviews, on social media – but we don't act. This lies at the root of our problem. How could we have wasted more than 50 years talking about the Climate Crisis but done so little to manage it? How can we accept homeless people on our streets when we know how to eliminate homelessness? How can we tolerate those hundreds of millions going to bed hungry while we waste food?

A closer look at some of the challenges that have threatened us for several decades shows that we seem unable to act decisively and collaboratively until we already suffer from significant damage.

Take climate change, for example; the first Climate Conference took place more than 50 years ago. We have known for a very long time what we need to do to avert a catastrophic future. Yet only now, when we start to suffer from the impact of climate change, are we taking some – but still only moderate – action. Understanding that we face a major challenge and knowing what needs to be done does not seem enough to jolt us into concerted, collaborative action.

As a species, we seem to have many characteristics, beliefs and values that work against us taking concerted action on the major issues. These include, for instance, a belief in staunch individualism over communalism and competition over collaboration. Also, our insistence on short-sighted action that addresses immediate and short-term goals and our reliance upon adopting the financially cheapest solution.

Some years ago, Prof Homer-Dixon coined the term *'Ingenuity Gap,'* suggesting that our problems have reached a level of complexity beyond our comprehension.[1] He may well be right. But we feel we are suffering even more from a *'Commitment Gap'*: We are happy to talk, fund, and carry out research, demonstrate, stage conferences, write books and papers – but we are not truly committed to taking decisive action until we are directly threatened in a tangible and personal way. Then we act. We put off doing something that may prevent extremely unfortunate outcomes until we must take action.

Yet, despite our apparent inability to address the major global crises we are facing, there is evidence that we can tackle the refugee crisis, the lack of organ donations, alcohol abuse, or political stand-offs at a local level. Why is this so? Is it the scale of a problem that freezes our brain cells? A closer look at these success stories reveals a different story: The approach taken to address these challenges was invariably aligned with how the human mind works. Is this the secret sauce? Are we unable to address our major global challenges because our attempts to address them are misaligned with the natural design of our mind?

Some of these success stories date back to a time when we had few insights into the workings of the human mind. Maybe the approach employed was based on intuition; perhaps it was luck. Fortunately, today, advances in cognitive science, behavioural economics, neuromarketing, and related areas deliver deep insights into what is driving people's actions.

Given our collective focus on money, we use these insights primarily for commercial gain: To get consumers to spend more or employees to be productive and loyal; to influence court outcomes or get people to buy into scams. But the very same insights can also help us address some of the massive challenges we are collectively facing. Understanding what drives people to act and act in certain ways might lift our chances of success. Would this provide the insights we need to close the Commitment Gap? Would we be more successful if we aligned our approach with the way our brains work?

We also need to acknowledge the source of the issues and problems we are experiencing in our world today. We speak about environmental or climate issues and in doing so we place the problems within the climate and the environment. We talk of social justice and race issues and again compartmentalize these externally to human beings. We give them a life of their own that we peer into. However, the issues above, as with many others, are issues of human behaviour. There is no problem with the climate or the environment; the problems are with the way we behave and the consequences upon the climate, the environment, people of colour, etc. We must take responsibility for our actions if we are to change these.

This book results from a strange mix of doubt, fear, and euphoria. Doubt that what we are doing today is not working; fear that we may already be too late; and the euphoria that comes from hoping that we can still make it happen by taking an informed approach – not informed by facts and figures we have already analysed *ad nauseam* with little action to follow but informed by what is driving human behaviour.

Here is what you will find in this book: Our first step is to make sense of our collective inability to address major, global challenges, even once we accept that they will eventually lead to catastrophic outcomes. What is odd about our behaviour is that we see the same pattern across very different challenges. This suggests that what we are observing is not challenge-specific but simply reflects human behaviour. If this is the case, we need to understand what drives our behaviour so we can shape it.

A useful first step may be to review examples where our efforts have successfully addressed challenges. They may not be global but rather local, but they nevertheless present significant challenges that have remained unsolved in most parts of the world. The cases we present in Chapter 2 cover different periods and countries. The obvious questions are: What can we learn from these case examples? Why did these local approaches succeed when it seems near-impossible to get collective action to address major global challenges? The answer may not be all that surprising: The common thread running through all these success stories is the alignment of the approach taken with how the human mind works.

This naturally leads us to Chapter 3 – exploring how the human mind works. More specifically, we focus on key drivers of behaviour. The important point is that it is fruitless to talk about human nature being 'good' or 'bad' as value judgments do not help us make progress. Instead, we need to understand why humans act the way they do – and why they don't take decisive action once a major challenge has been identified. Our investigation provides the scientific foundation validating the Commitment Gap model we introduce in Chapter 1.

So far, we have explored how we got from evolution, i.e., an improvement in our quality of life, to devolution, i.e., an apparent inability to deal with major challenges even though we understand that they will eventually reduce our quality of life or even shorten it. It is now time to move on to some practical considerations by asking how to trigger decisive action. In other words, we are getting ready for our revolution, which takes us to the second part of the book.

Naturally, we don't want to turn this into some theoretical treatise or navel-gazing exercise. The safest way to avoid this is to anchor our conclusions in real-world examples (outlined in Chapter 4). With this in mind, we are reviewing four major, but very different challenges, namely homelessness, the refugee crisis, food insecurity, and gender discrimination. In each case, we use some data to understand the scope and scale of the challenge and identify broad directions designed to trigger decisive action that can seriously impact the problem we are facing.

These broad directions are, hopefully, a sound starting point. But as the saying goes, the devil lies in the details. With this in mind, we present a collection of specific guidelines and tools in Chapter 5, providing the reader with a range of options on how to address our major, global challenges.

Conclusions are left for Chapter 6. Our intent is to illustrate the journey from evolution to devolution and, finally, revolution. The first two steps are evolving and, quite rapidly, with some of the major challenges we face. We are moving from a period of growth, increased prosperity, and unparalleled quality of life into a time of devolution, which sees us lose much ground on these fronts. To address this situation, we first need to understand why this is happening. We are supposedly the most intelligent species on earth, yet we are the very cause of our own demise. There is no doubt that we have a Commitment Gap. We understand the problem, we know how to address it, but we fail to take decisive action.

But how can we change this situation? The most important step we need to take is understanding what is driving our destructive behaviour and developing solutions aligned with how our mind works. We won't stop our decline and return to an evolutionary phase overnight. Still, there is hope that we can slow down our decline and transform the world we have designed fast enough to avoid a catastrophic outcome.

At this stage, we must admit to a major shortcoming: we have largely ignored the much more immediate problems faced by the citizens of many developing countries. There is no point suggesting grand solutions when you must fight just to live another day. Instead, the developed world needs to push ahead and invest in solutions that will also help reduce the plight of people fighting with poverty, hunger, the lack of healthcare or social services, and no future to aspire to.

Let us finish this brief overview with a personal observation: Just about all the major, global challenges that threaten our quality of life – if not our actual life – have been man-made (in this case, the use of the male gender is appropriate!). Think about the refugee crisis, homelessness, food insecurity, gender and age discrimination, the Climate Crisis, and more. Today, we have an abundance of technologies, time, and resources to address these challenges. But we are too busy to do so. Those who are lucky to enjoy a good life are too busy entertaining themselves and enjoying the privileges our economic progress has delivered. Others are too busy trying to survive. If the former group does not start to take concerted action to address our shared challenges, we will see more people ending up in the latter category – and evolution will firmly turn to devolution.

This is the tragedy of the 21st century: The use of our technological brilliance to enrich our lives in what are, in the end, meaningless ways. We pursue space travel or the development of metaverses, of new entertainment options and apps that make us feel great and take our focus off the real challenges we need to address – if not for our benefit, then for generations to come.

We are certainly not advocating a monk-like existence. There is nothing wrong with enjoying life and living it to the fullest – but we also need to spend some of our time, energy, ingenuity, and resources addressing the major challenges that will destroy the quality of life for many – and life for some. It is all very well to look down from space and talk about how beautiful and fragile the world looks – but should we invest in space travel, improved social media concepts, the metaverse, enhanced video games, and other entertainment opportunities, or should we put our energy behind an effort to protect our quality of life for our future and that of generations to come? And should we not be responsible for helping others, less fortunate than us, to improve their quality of life?

We look at our grandchildren and feel lucky that we are not their age. They will not have the fantastic and easy life we had an opportunity to live. Unless we act today, they will face crisis after crisis, live with uncertainty, and find their well-being and even their lives at risk. Time is of the essence. We can avert the worst, but not if we continue to just talk about it rather than take decisive, collective action. We leave these thoughts with you and hope you will be an active contributor to closing our collective Commitment Gap!

Following the story...

One of the authors is a leading expert on, and the originator of, the Declarative Mapping Sentences (DMS), providing us with an opportunity to structure our presentation and their multiple components. The DMS is drawn from the Declarative Mapping Approach (DMA) which is a whole way of thinking about the gathering and understanding of research. The DMS is employed in social science research and has since been used to guide research into many phenomena. Simply put, the DMS is a way of offering a road map for understanding a complex area of interest. The maps produced with a DMS, rather than being pictorial or image-based, take the form of a sentence of written language.

In a DMS the sub-aspects of a phenomena, event, or issue (which are called facets) are combined into a sentence and are joined together in such a way as to suggest the relationships between these facets. Each of the facets represents a major sub-domain of the area of interest and each facet is broken down in its own subcomponents called elements. An example of a DMS will assist readers to better understand the DMS. We may have used any area of human behaviour, but let us consider how we experience the place where we work. In this situation a DMS may be written as follows (Figure 0.1).

FIGURE 0.1   EXAMPLE DECLARATIVE MAPPING SENTENCE FOR A PERSON'S EXPERIENCE OF THEIR WORKPLACE

A person experiences their workplace in terms of the:

Physical

cramped
spacious

physical aspects of the place, which:

Social

facilitate
neither facilitate nor retard
retard

social interactions, and which they perceive to be:

Aesthetic

aesthetically pleasing
aesthetically neutral
aesthetically unpleasant

and which:

Functional

enhances productivity
has no effect on productivity
retards productivity

We use this example of a DMS that is not related to the Commitment Gap in order to illustrate the generic nature of this approach. Later in the book we directly address the Commitment Gap using the DMA.

All mapping sentences are read from their start to end, as with an ordinary English language sentence. However, when the reader gets to a facet (emboldened and italicized) the facet is not read but rather one of the elements (hyphenated and indented) is selected from each facet and read in the facet name's place. The way the facets are arranged in the DMS reflects the meaning present in the domain of interest and the way in which the facets interact. The DMS also acts as a reminder to someone who is interested in the domain covered by the DMS that it is not a simple domain that may be understood by trying to assess this by concentrating exclusively on a single aspect (facet). For instance, in the above example if someone is trying to understand a person's experiences in a workplace they cannot do this simply by looking at how functional the place's design is (for example). Rather, all aspects (or facets) of experience must be considered.

In this book, we use the DMA as a methodology to guide and inform our understanding and the DMS to structure and analyse our findings and discussions. It is important to note that the DMA and DMS, whilst foundational to this book, are not the primary focus, but rather methodological tools, used to unveil the theoretical premise and analysis behind our discourse. For sake of reading continuity, we do not present DMSs for every example mentioned, but please know that the DMA and DMS has been employed in our sense-making process.

Target audience

To conclude this preface, it should be noted that the writing and content in this book is aimed at an academic audience and, most specifically, at those interested in or teaching and studying research methods.

Note

1 Homer-Dixon, Thomas, *The Ingenuity Gap: Facing the Economic, Environmental, and Other Challenges of an Increasingly Complex and Unpredictable World*, Knopf, New York, 2000.

1

THE COMMITMENT GAP

The first climate change conference took place more than 50 years ago, but little progress has been made since then. The refugee problem has been building for decades, fueled largely by military interventions and civil unrest, but little has been done to manage this challenge. Income inequality has been a major concern for decades, yet no decisive action has been taken. Gender and age discrimination are rife in most parts of the world, but little has been done to address them. Homelessness is a problem that could be addressed effectively and quickly, but most of the world's cities have failed to act.

We have made little progress with addressing the key challenges that threaten our quality of life and, for some, even life itself. We are, of course, aware of these issues, have tangible evidence of their existence, are exposed to them in the media, conduct research studies, formulate policies, and hold talkfests to explore them. But decisive, collective action is rare. Indeed, we are unlikely to take such action even once there is irrefutable evidence that we will be severely affected by an evolving crisis.

Humankind's response to major challenges seems to invariably follow the same pattern, even though the problems differ so greatly. It clearly makes no sense to keep doing the same thing while expecting a different outcome. We need to change how we address our major challenges. First, however, we need to understand how we developed such an ineffective response pattern, and what that means for finding new ways to shape our behavior.

DOI: 10.4324/9781003477167-1

Humans claim to be the most intelligent species on earth. Therefore, one might expect us to decisively address major challenges rather than just watch them evolve into major catastrophic events. But the evidence shows otherwise.

We seem to have a standard response to major challenges: we move through several stages before we are ready to take decisive action. This pattern is our standard response because it reflects what we commonly call 'human nature', our built-in propensity to ignore any threat that is complex but which does not yet present an immediate. and personal danger. And because these threats tend to grow incrementally, we gradually get used to them, dulling our responses in the process.

As a result, we focus on dealing with the smaller issues caused by the evolving challenge, rather than considering the challenge itself. We establish refugee camps because we must put the refugees somewhere; we run some ad campaigns promoting gender equality to satisfy anyone feeling strongly about this issue; we offer limited government subsidies for clean energy projects so that we can point to our commitment to reducing emissions.

When our leaders do something, it's mainly to be *seen* to do something, rather than any serious attempt to address the core challenge decisively and effectively. They have climate talkfests so they can proudly point to some bilateral or multilateral agreement, but little changes. We, the citizens of this planet, are meanwhile simply looking for some assurance that we will be okay, rather than demanding decisive action. We want to be told that there's nothing to worry about, that technology, innovation, government policies, time, or some other magic will make our problems go away – and our political leaders are happy to deliver such promises.[1]

To start with, we need to understand the stages we typically go through when facing a major challenge.

The Hackett Model

We may like to think that we are unique individuals who find our own unique path in life. To some extent this may well be true, but it is a fact that we largely share our responses to danger and, by extension, to major challenges.

For example, if we consider the experiences in our own lives, most of us may be able to think about a road junction or some practice at work about which we have always had doubts. Every time you approach the road junction you say to yourself, *"This is just an accident waiting to happen!"* Or there may be some equipment at work that you use and think, *"If I was not careful this could be really nasty!"* At that point you most likely say to yourself, *"Surely, I can't be the only person who has noticed this"* and, of course, you are not. However, even in the face of obvious danger we find it difficult to act until a major accident has occurred.

Similarly, when addressing individuals who have different points of view to our own, we tend to follow a well-trodden path. According to Kathryn Schulz, we progress through three stages of labelling people with differing views to ourselves (Kathryn Schultz et al., 2010).

To start with, when we encounter someone with an opposing view to our own, we feel that they are simply ignorant. We assume that the person is simply unaware of the facts as we understand them, and we believe that once we have informed them of the necessary information, they will agree with us. At this point we may share information with them and be horrified that even once we have done this, they still hold an opposing view.

We then progress to the second stage which goes something along the lines of, *"If I have given you all this information and you still disagree with me, you must be really stupid"*. This is an *ad hominem* attack which is unfortunate and escalates the possibility of conflict between the parties involved.

However, we may find out that the person who disagrees with us is actually pretty smart and not intellectually deficient, which leads to things getting even worse in stage three, where we reason, *"I have given you all the information you need to understand the issue and to agree with us and you are an intelligent individual – therefore, you are intentionally distorting the information for your own malicious purposes"*. In essence, we label the person who disagrees with us as evil.

We also tend to have a common response when being confronted with major challenges. One of the authors developed a model that shows a typical response in the face of environmental challenges (Hackett, P.M.W., 1995). In his research he looked at an individual's support for, commitment to and actions associated with a series of pro-environmental or conservation measures and discovered a four-stage process.

He found that once a person became aware of an environmental issue (stage 1) in order to do anything about this they next had to believe that this issue was important (stage 2). In stage 3, the person had to be convinced that it was possible to effectively address the specific issue in order that they actively supported this pro-environmental activity. Finally, the person had to believe that what they themselves could do would make a difference (stage 4). From this research, it appeared that if the four stages could be progressed through and positive states achieved at all stages then they were more likely to commit time or money, or to become actively involved in other ways in efforts to address this specific issue. It should also be noted that after stage 4 the individual may then revisit stage 1, which will have been potentially altered by the process, and repeat their way through the remaining three stages of the model. This process could potentially be repeated in a never-ending cycle.

These findings (and the model) may also relate to other forms of human behavior regarding environmental and social (eco-social) issues. One example of this more broad process may be understood through the use of a declarative mapping sentence (Gordley-Smith and Hackett, 2023; Hackett, 2014, 2018, 2019, 2020, 2021; Hackett and Lustig 2021; Lustig and Hackett, 2020a, 2020b) (Figure 1.1).

FIGURE 1.1   THE HACKETT MODEL: A DECLARATIVE MAPPING SENTENCE (DMS) REPRESENTATION

A typical response in the face of a major issue may be dependent upon a person's level of:

Stage 1

awareness
unawareness

that the challenge exists, which then leads to belief that the challenge is:

Stage 2

important
not important

and that the person may also believe that their actions may be:

Stage 3

effective
ineffective

and that there is a/an:

Stage 4

possibility of Success
impossibility of success

which then leads the person to behave in a certain way, and to return to stage 1 and to develop awareness (or not), of the now modified major issue.

This DMS may be adjusted and expanded upon for specific and complex environmental, social and other challenges and issues. Later in this book we will employ this particular DMS in respect of refugee challenges.

Our Emotional Reaction to Major Challenges

FIGURE 1.2   TYPICAL STAGES WE GO THROUGH WHEN FACING A MAJOR CHALLENGE

Building on the Hackett model, we have identified several stages defined largely by the way people feel about a major challenge, ranging from being delighted to living in denial, being racked by doubt to feeling desperate, leading finally to decisive action.

Positive

Negative

This model as represented in a DMS may be seen below. Through the use of this DMS, one may be able to more clearly extrapolate the connection between the Hackett model and the emotional stages one typically goes through when facing a major challenge (Figure 1.3).

FIGURE 1.3   THE DECLARATIVE MAPPING SENTENCE FOR THE TYPICAL STAGES A PERSON GOES THROUGH WHEN FACING A MAJOR CHALLENGE

The initial response a person goes through when they are faced by a major challenge is:

Stage 1

increased delight (enthusiasm)
decreased delight (enthusiasm)

along with:

Stage 2

increased denial
decreased denial

requiring the person believe that their actions may:

Stage 3

increase doubt
decrease doubt

and that there is a/an:

Stage 4

increased despair
decreased despair

associated with:

Stage 5
increased decisive action
decreased decisive action

Let's explore each stage in greater depth:

Delight

After the Second World War, the western world enjoyed approximately six decades of economic growth and wellbeing. Many of us enjoyed an elevated lifestyle without having to worry about food or water security, financial disasters, wars, or other calamities. We enjoyed the benefits of new technologies, meaning that every year we could look forward to an even better 12 months ahead. This was a time of prosperity and a time that could have been used to prepare the world for the many challenges we would eventually face.

Were we aware of these challenges? Let us remind you that the first climate conference took place more than five decades ago, that the refugee crisis was as acute as ever after the Second World War, that food insecurity has always been an issue, and that gender equality has been absent in most parts of the world throughout written history. Yes, there was an awareness that pressing problems were building up.

Unfortunately, at critical inflection points when our leaders could have changed how we address major challenges, they had different ideas. They chose to advance their own competitive agenda rather than benefit the world at large.

For example, the world could have benefitted enormously from an international organization that was well-funded, independent, and had some enforcement powers. Instead, we got the United Nations (UN). The UN's permanent members – the United States, China, Russia, France, and the United Kingdom – have the right of veto, allowing them to remove from the global agenda any items that did not please them. Did anyone ever believe the UN could play a decisive role in building a better world when the world's power brokers could kill off any initiatives that did not suit them? Clearly, the United Nations was designed to serve the interests of the most powerful nations rather than serve the world and indeed some may claim that it is dominated by the powerful Western nations.

The US ensured that the global financial system was under its control, allowing it to impose sanctions on any third party that displeased it. It has made frequent use of this power and, in doing so, has contributed significantly to hunger and desperation in countries that lack resources and rely on trade and financial support from friendly nations.

In brief, there were windows of opportunity to establish international organizations that could address major challenges and establish positive relationships between nations. Of course, the outcome would not have been perfect, but it could have been far superior to what we got.

Meanwhile, citizens benefiting from the decades of growth after the Second World War were too busy enjoying the rewards economic growth brought, the fruits of technological progress, and the opening of the world to trade and tourism, to take too much notice of the major challenges that were evolving. It's somewhat ironic that so many people today feel overwhelmed by the demands life makes on them, ignoring that social media, gaming, and other entertainment options are a choice they make, not a duty they are burdened with. Whatever the reasons, at present there seems to be little appetite for addressing our collective challenges.

Denial

As a challenge evolves, it reaches a stage where it becomes impossible to ignore. But until it affects us personally and immediately, we prefer to live in denial. Even when we are exposed to reports that thousands of people are suffering or have died, we may feel sympathy, we may even be outraged, but our emotional involvement is typically superficial and quickly recedes.

Ignoring the problem is made easier by our tendency to favour opinions and information from websites, news programmes, and social media posts that reinforce our belief that we will be okay. The climate crisis – something the next generation will have to pay attention to! Gender discrimination – don't forget how much progress we have already made! Pollution – sure, industry needs to lift its game! Refugees – I feel bad about them, but at least we are keeping them safe in camps!

The COVID-19 pandemic is a good example of this. It caused the deaths of well over 6 million people, and experts agree that this number is severely understated because many countries do not keep accurate statistics (Salvador Rizzo and Nirappi, 2022). Everybody – globally – must get vaccinated to slow down the pandemic's progress and avoid the development of new, more potent strains. But many people are unwilling to be vaccinated, citing their right to choose. In other words, they see their comfort as more important than the lives of others. Of course, it is not possible to know how the vaccination rates may have been different in countries such as the US if a different socio-political system had been in place when the pandemic struck. It does seem reasonable to assume, however, that the neo-liberal free-market capitalist system, which emphasizes the individual and him or her own wants, did little to instill a sense of communal responsibility within the country. Unfortunately, such a social design is present in much of the world and has shaped how we dealt with the pandemic.

On a larger scale, we saw the wealthy nations hoarding vaccines rather than supporting those nations that do not have the purchasing power to acquire the vaccines they need. (To be fair, some of the blame must go to Pfizer, the pharmaceutical firm that allegedly asked wealthy nations to sign a contract restricting the export of vaccines.)

We are, unfortunately, very good at rationalizing our decisions and our behaviour. For instance, when we go to buy a car, we may arm ourselves with information from the internet about performance and other users' opinions. We may also narrow down the makes and models in which we are interested. These are all sensible choice practices (although the opinions of other users may be rather biased and subjective). However, when we get to the show-room, we may be swayed by the salesperson or special offers and may select a totally different car to the one we planned to buy based upon the information we had gathered. Yet, we can convince ourselves that we rationally selected the best model for our needs.

Unfortunately, we also use this mental capacity when we don't want to do something and are seeking a rational reason for our lack of action, rather than admitting that we are simply too self-centred, devious, or lazy. This is where confirmation bias comes into play, which is the phenomenon that as individuals we find it comforting to hold beliefs and attitudes that are in harmony with each other: that is that they are consonant. This is shown in the cluster of beliefs regarding a variety of different policies that political parties embody.

What we mean by this is that someone who holds left-wing views about policing is quite likely to hold left-wing views about other areas of life such as education, social services, and so on. The same is true of right-wing, centrist and all other types of beliefs. Holding consistent views is a form of time and energy saving short-cut that we employ to limit our cognitive effort, i.e., the need to think extensively about every issue confronting us or every decision we need to make. This behaviour reinforces our tendency to be self-centred and inflexible.

At the same time, we expect our leaders to assure us that we will be okay and, while we know that they are often not honest, we believe whatever they say when it suits us. After all, 'bad things' are always happening somewhere in the world; we just don't want to hear that these events could impact us, personally, in a material way.

Doubt

Doubt develops when people we know or who are in a similar position to our own are affected by an evolving crisis. For example, more and more farmers in the United States Midwest are allegedly starting to doubt the long-held belief that climate change is a hoax. As irregular weather patterns, major drought and floods destroy their crops, doubt starts to seep into their minds. People living in coastal areas that have experienced significant and repeated flooding start to believe that the ocean is rising and that this will threaten their homes and livelihoods. Legislation and campaigns that target gender discrimination and violence against women suddenly make the problem personal: Everyone can get caught! No one can hide!

When doubt sets in, stress is never far away. Unfortunately, the stress response that helped humankind to survive in a hostile natural environment is designed for acute stress: our brain releases hormones that damp down the immune system, increase alertness and quicken the heart rate in order to create the best possible energy-bristling state (Peter Steidl, 2017). But this response doesn't help us to deal with the chronic stress we experience when we worry about the impact a major challenge may have on us and those closest to us. The constantly heightened arousal triggered by chronic stress impedes our ability to make sound decisions. Some people rush into impulsive decisions, while others find themselves unable to make any decisions at all.

At the same time, we see commercial enterprises capitalize on a stressed-out populace. Demand for self-help apps, programmes, and advice promising an escape from the world we live in surge as people feel pressure and stress, and all they want is to get to a better place. Of course, there is nothing wrong with an occasional escape to calm and balance. But when our efforts are entirely focused on looking after ourselves rather than addressing our collective challenges, we become part of the problem. We need people to feel better because we have started to address our major challenges constructively, not because they have escaped reality altogether.

Not surprisingly, decisive action is a most unlikely outcome at this stage. But at least the seeds of doubt have been sown.

Despair

Despair sets in when we are personally affected or believe that we won't escape the impact wrought by a major catastrophic development. Suddenly, it is about us. The danger is personal and imminent, and our brain goes into overdrive and pushes us to seek a way forward.

However, the human brain is hardwired to *focus our attention on our very own personal situation* rather than on how we could collectively address the challenge we face. It's about everyone for themselves – and today's challenges cannot be addressed by individuals just looking after their own wellbeing.

To add to the problem, our brains are also hardwired to take shortcuts, especially when we feel under threat. This survival mechanism, another relic from the times when we lived with constant physical threat, served us well when we had to react quickly rather than think things through. Unfortunately, it also makes us more likely believe the glib claims of politicians who promise to solve our problems for us. We don't care that there are no sound policies, no budget analyses, or details underpinning their promises. Believing that someone will magically solve our problems is enticing and believing a solution is just around the corner will trigger a dopamine hit, allowing us to feel better instantly.[2] No wonder politicians spend so much money conducting surveys to find out what people worry about – so they can blithely promise to address these problems once they are elected.

Decisive Action

When the situation gets desperate, one hardwired brain circuit may help: our hardwired drive to belong, which encouraged humans to join forces to increase the likelihood of survival in a hostile natural environment. As David Foster Wallace said: *'Nothing brings you together like a common enemy'* (David Foster Wallace, 2006). When we reach this point, we start to consider a collaborative effort that transcends social, political, philosophical, religious, and ethnic lines. Driven by desperation, we are finally ready to address our challenge collectively. But because we have wasted so much time getting to this point, there are likely no simple solutions left – in fact, there may simply be no solutions left at all. This is shown in many political confrontations that often end in armed conflict.

FIGURE 1.4   WHERE ARE WE ON OUR JOURNEY TOWARDS DECISIVE ACTION WITH RESPECT TO MAJOR CHALLENGES WE FACE TODAY?

Figure 1.4 shows how far we have progressed on the 'doomsday curve' for several major challenges. Obviously, the situation differs by country and even within countries. You may not agree with where we have placed each challenge, but if you agree with the overall trajectory, you are ready to ask yourself what we need to do to overcome our in-built drive to destroy ourselves (Figure 1.5).

Positive

Negative

FIGURE 1.5   A DMS EXPLORATION OF WHERE WE ARE ON OUR JOURNEY TOWARDS DECISIVE ACTION WITH RESPECT TO MAJOR CHALLENGES THAT WE FACE TODAY (FIGURE 1.4)

The journey towards decisive action with respect to major challenges we face today may be illustrated as showing an:

Stage 1a

- increased delight (enthusiasm)
- decreased delight (enthusiasm)

toward addressing the:

Stage 1b

- homelessness
- refugee crisis
- gender discrimination
- climate crisis
- Ukraine War

along with an:

Stage 2a

- increased denial
- decreased denial

in respect of the:

Stage 2b

- homelessness
- refugee crisis
- gender discrimination
- climate crisis
- Ukraine War

requiring the person believe that their actions may:

Stage 3a

- increase doubt
- decrease doubt

towards the:

Stage 3b

- homelessness
- refugee crisis
- gender discrimination
- climate crisis
- Ukraine War

and that there is a/an:

Stage 4a

- increased despair
- decreased despair

in respect of:

Stage 4b

- homelessness
- refugee crisis
- gender discrimination
- climate crisis
- Ukraine War

along with,

Stage 5

- increased decisive action
- decreased decisive action

How long each of the five stages – Delight, Denial, Doubt, Despair, and Decisive Action – takes depends on the type of challenge we are facing. For example, climate change has been on the agenda for more than 50 years. We are finally getting ready to move from talking about what we could do and addressing a few fringe issues to taking some sort of collective, decisive action, though we are still facing setbacks because many governments value today's economic gains more than a future inhabitable earth, as evidenced by Australia continuing to commission coal mines, Brazil allowing for the further deforestation of the Amazon, and Canada mining tar sands.

The Commitment Gap

Almost twenty years ago, Professor Thomas Homer-Dixon, the Canadian academic, author, and consultant, coined the term 'Ingenuity Gap' to describe the distance between the demands the challenges we face pose and our ability to address them. We believe there is another, even more damaging gap: The gap between knowing what to do and taking decisive collective action. We know what we need to do, but we are not prepared to act. We have known about a wide range of measures that would have allowed us to manage the climate crisis for decades – but we have not taken decisive action. Similarly, we know that some cities have eliminated homelessness, but we don't see other cities using their proven strategies.

Right now, we could act to de-escalate the refugee crisis, eradicate homelessness, limit global warming, or address impending water and food shortages.

But we won't.

We are also hardwired to invite tragedy before we are willing to act. We are programmed to chase desire and delight for as long as possible and, when we can't avoid being confronted with future risks, we move into denial. We are conditioned to look after ourselves first and foremost. This has triggered a rise in libertarian and individualistic politics and, as we remarked earlier, this seems to take us away from collective feelings and actions. Only when we face personal and immediate threats will we finally start to have serious doubts and reach the point of despair. Once we get there, we are ready for decisive action, but the day will come when we arrive at this point far too late to avoid our own endgame.

However, all is not lost. We may be failing to deal successfully with the major global challenges we are facing. But there have been isolated cases of success: Major problems, seemingly unsolvable worldwide, have been addressed effectively at the local level. While this may happen all too rarely, it is possible that these successes stem from taking a particular approach to problem solving rather than being random events. And if that is so, we should be able to learn from these isolated cases. But will we do this, and will be quick enough?

But before turning our attention on these local success stories, let us illustrate the Commitment Gap by reviewing how we approached the evolving climate crisis.

Illustrating the Commitment Gap: The Climate Crisis

"One of the basic causes for all the trouble in the world today is that people talk too much and think too little." Margaret Chase Smith

Let's go through the history of the climate crisis step-by-step, albeit in a highly compressed form. In what follows we are drawing on the work of Alice Bell (Alice Bell, 2021). We highly recommend her book to anyone who is interested to learn about climate change and how the world failed to deal with this challenge.

Delight

The second half of the last century was arguably the most prosperous time ever experienced in the Western world. The United States flourished as it reaped the benefits of its victory in the Second World War. Rebuilding Europe, Japan, and other devastated areas, the subsequent globalization of trade, and technological developments often based on advances made by the defence industry, all fuelled the growth engine that propelled the world to unimagined riches, secured a prosperous life for many, and delivered year-on-year growth with just a few bumps on the way.

Despite some warnings, climate change was not a significant concern during this period. People were too busy rebuilding their lives and growing their wealth to be concerned about something they only vaguely – and very occasionally – heard about. The distrust of claims for climate change were exacerbated by the scientific claims made in the middle-late 1900's that we were heading for another ice age. The apparent reversal from a freezing to an ever-warmer world was often used to justify scepticism in the claims of science.

Denial: 1959 to 1989

As more evidence showed the damage human emissions caused, denial was the most common response.

Naturally, parties that benefitted from the widespread use of fossil fuels were only too happy to plant the seeds of doubt. In 1959, for example, Shell published an article in the *New Scientist* in which it stated that that carbon dioxide accounted for only 0.03 percent of the atmosphere, most of which was from natural sources.[3] If you were still concerned, there was an assurance that the oceans and plants dealt naturally with carbon dioxide. In other words: nothing to worry about!

Climate critics offered another reason for plausible denial: maybe there was global warming, but – as the *Times* reported – this would be followed by a period of global cooling. Notably, most parties – both for and against climate change – agreed that it would take a considerable time for the effects of any climate change to be felt. There was no immediate danger. In 1969, a series of BBC lectures warned of the dangers of warming oceans and the melting polar ice caps – but told its concerned audiences that all this would happen 'sometime' in the future (BBC, 2011). There was no immediate threat.

The first Earth Day, held in April 1970, was a highly successful event with 1,500 colleges and 10,000 schools staging teach-ins, events in a wide range of public places, and demonstrations at business and government offices. It was estimated that up to 10 percent of the US population was somehow involved (earthday.com, n.d.).

Earth Day targeted many issues, from sewerage to pollution, the plundering of limited resources by the United States, and emissions, to name just a few. The emission issue focused primarily on pollution and, in particular, smog, as this was something people in some cities were experiencing directly, in contrast to climate change, which was, at that time, a somewhat intangible concept that might present a danger in the distant future.

During this period, scientific evidence about climate change was becoming overwhelming and the latest research was presented at a 1972 UN conference (un.org, 1972). This suggested that it might take some 30–50 years of work to avoid the worst consequences of the phenomenon. In other words, urgent action was required, despite the lack of impact at the time. Again, however, there was a diverse agenda covering deforestation, endangered species, pollution from industrial or military chemicals, radiation, the population explosion, and other issues. Don't get us wrong: all these issues were important, and some, such as deforestation, were directly related to climate change. We simply note that it was impossible to get a strong focus on climate change with so many competing challenges.[4]

It is also worth noting again that this relates to the second stage of the commitment to pro-environmental activities during which a person asks if the issue is of importance to them, and to the third stage, when they ask about the effectiveness of any form of fix for the problem. Finally, in stage 4, they assess the extent to which they feel their own actions can make a significant effect on the issue.

Support of some kind came from an unexpected quarter – the CIA. Concerned about climate change's possible geopolitical impact, the CIA produced a study that warned of the political unrest and mass migration that would result from food shortages (CIA, 1974). Unfortunately, the CIA report talked about global cooling as well as warming.

Regardless, the oil crisis was a more immediate challenge for many Americans at the time.

The fossil fuel industry started its lobbying and misinformation campaign in all seriousness during the last quarter of the 20th century. Their strategy at this time was to sow doubt: we can't be sure yet as there is not enough scientific evidence to support global warming or evidence linking it to human activity. Doubt is an effective strategy as the audience is receptive, seeking assurance that they have nothing to worry about.

The scientists went on with their conferences and workshops, yet their impact was limited without political and public support and in the face of a hostile fossil industry lobby. This period saw the World Climate Conference in Geneva in 1979, and a new Carbon Dioxide Assessment Committee established by the National Academy of Science in 1980. The White House made it clear to the Academy that it did not want anything announced that would alarm the public, and the report stated that there was reason for caution, not panic. This led to the *Wall Street Journal* reporting that '*A panel of scientists has some advice for people worried about the much-publicised warming of the Earth's climate. You can cope.'*

By 1981, less than 40 percent of Americans had recalled hearing or reading about the greenhouse effect. And for people who had heard about the issue, it had generally been presented as something far off, or that would be sorted out soon enough.

Transition from Denial to Doubt: 1987–1989

A report on the environment was presented to the UN General Assembly in 1987, which promoted the idea of sustainable development, i.e., meeting the needs of the present (especially the world's poor) without compromising the lives of future generations (World Commission on Environment and Development, n.d.). It was decided that a new UN Conference on Environment and Development would be held in 1992 in Rio de Janeiro; this was given the name the Earth Summit. Note that the summit was planned to take place five years later – there was no sense of urgency!

Public awareness of climate change increased over the years. In 1988, a survey showed that 58 percent of Americans recalled having either heard or read about the greenhouse effect (in 1981, the figure had been 38 percent). Other polls showed that most Americans believed the greenhouse effect was either extremely or very serious and worried either a fair amount or a great deal about global warming. In 1989 *Time* magazine, instead of selecting a person of the year as was its custom, declared 'the Endangered Earth' to be the Planet of the Year (TIME, n.d.).

But politicians, the oil industry, and even some scientists continued to muddy the waters. With her scientific background, the then UK premier Margaret Thatcher recognized the dangers of climate change, but at the same time she suggested that this was not something the average citizen needed to worry about or that they should blame industry for. In 1989, speaking at the UN, she said 'We must resist the simplistic tendency to blame modern multinational industry for the damage which is being done to the environment. It is industry which will develop safe alternative chemicals for refrigerators and air-conditioning. It is industry which will devise biodegradable plastics. It is industry which will find the means to treat pollutants and make nuclear waste safe – and many companies as you know already have massive research programmes' (margaretthatcher.org, 1989).

Similarly, on the campaign trail for the 1988 election, US Republican presidential candidate George Bush Sr declared: 'Those who think we are powerless to do anything about the greenhouse effect forget the White House effect.' The message was clear: global warming is an issue, but the government, technology, industry, and science will deal with it. There was no reason for the average citizen to worry about this.

Doubt

In 1990, the George G Marshall Institute, founded by three scientists to counteract the Union of Concerned Scientists, published the report *Global Warming: What Does the Science Tell Us?* (Robert Jastrow et al., 1990). They argued that the Sun had been responsible for the slight warming of the Earth which had occurred in the previous century, and that when its natural variation calmed down, that would balance out any future greenhouse warming. This appealed to the White House, particularly the then Chief of Staff John Sununu, who was far more concerned about any negative economic impact than the future of the planet. He banned the terms 'climate change' and 'Global Warming' from any White House communications.

In April 1990, an internal White House memo on talking to the press about climate change was leaked. This confirmed that the oil industry and the White House were actively seeding doubt. The memo included the advice that there was no point in arguing global warming was not happening, but 'to raise the many uncertainties that need to be better understood on this issue.'[5]

More conferences took place, and reports were issued, but nothing of substance changed. Let's jump forward to the World Climate Conference in Rio. The plan was to have a text for a UN Framework Convention on Climate Change (the UNFCCC) ready for heads of state to sign at the Earth Summit in June 1992. However, the US was not prepared to sign anything demanding emission cuts in the short term.

The Rio Earth Summit was a major media event. The UN's Secretary-General declared, 'Ultimately if we do nothing, then the storm will break on the heads of future generations. For them it will be too late.' The results were a number of loose agreements to 'stabilize greenhouse gas concentrations in the atmosphere at a level that would prevent dangerous anthropogenic interference with the climate system.' In other words, no decisive action was taken – again. As with other conferences and events, the only outcome was a promise to keep talking, not take decisive action. To facilitate the talking, the UN set up a new body – the UNFCCC – that would facilitate an Annual Conference of the Parties (i.e., countries), also known as COP.

And then the talkfest really started. Here is a list of COP meetings to date (UNFCCC, n.d.):

1995	Berlin, Germany
1996	Geneva, Switzerland
1997	Kyoto, Japan
1998	Buenos Aires, Argentina
1999	Bonn, Germany
2000	The Hague, Netherland
2001	Marrakech, Morocco
2002	New Delhi, India

2003	Milan, Italy
2004	Buenos Aires, Argentina
2005	Montreal, Canada
2006	Nairobi, Kenya
2007	Bali, Indonesia
2008	Poznan, Poland
2009	Copenhagen, Denmark
2010	Cancun, Mexico
2011	Durban, South Africa
2012	Doha, Qatar
2013	Warsaw, Poland
2014	Lima, Peru
2015	Paris, France
2016	Marrakech, Morocco
2017	Bonn, Germany
2018	Katowice, Poland
2019	Madrid, Spain
2020	Glasgow, Scotland
2021	Paris, France
2022	Sham El Sheikh, Egypt
2023	Dubai, UAE

If you think only a group of policy makers supported by leading experts was meeting in some of the most beautiful places on Earth, you are quite wrong. 70,000 people were accredited to attend COP 23 in Dubai, with some of them arriving by private jet at an event that is supposed to find a way to reduce emissions. It is dispiriting that with so many COP meetings, involving so many participants, so little has been achieved.

But let us now return to our chronological review:

In 2002, Republican consultant Fred Luntz advised the party's leaders:

Voters believe that there is no consensus about global warming within the scientific community. Should the public come to believe that the scientific issues are settled, their views about global warming will change accordingly. Therefore, you have to continue to make the lack of scientific certainty a primary issue in the debate.

(Frank Luntz, n.d.)

Kyoto had failed, but there was hope in 2009 that the Copenhagen COP talks would result in some firm commitments. The event was dysfunctional, and a side agreement which was made at the BRICS (Brazil, India, China, and South Africa) meeting between China and the US (which was not specific or binding in any way) was seen by other countries as clear sign of power politics that rendered the COP talks irrelevant. For many, the Copenhagen COP15 was a setback rather than a sprint forward.

Paris Climate talks in 2015 delivered an agreement to make a strong commitment to keep global warming to less than 2° above pre-industrial levels but to aspire to 1.5°. However, as it was left to each country to make their own plans, allowing them to start with timid, small steps that would somewhat magically grow in the future, it is difficult to see this agreement as a significant step forward.

Despair – We Are Not Getting Anywhere!

There are certainly pockets of despair surfacing. Some island nations already see their existence as being imminently threatened by rising sea levels; some regions are affected by extreme weather conditions, bushfires, or floods. They may not always attribute their misfortunes to climate change, but first-hand experiences have raised the level of concern. For a short period of time, there were numerous demonstrations, with the activist Greta Thunberg and the organization Extinction Rebellion getting the most media coverage (Wikipedia, n.d.).

But actions now seem to have quietened down, possible due to the interruption caused by Covid-19, or because we simply got used to these initiatives and the media does not feature events with which few people are engaging.

The same patterns can be observed with respect to the Global Week for Climate Action initiatives.

The September Global Week for Climate Action comprised of a series of strikes and events, held from the 20th to the 27th September 2019, designed to take place around the UN Climate Action Summit on the 23rd of September. Protests in some 4,500 locations were planned in 150 countries. We have seen nothing on a similar scale since then (Wikipedia, n.d.).

The pandemic curtailed the scope and scale of the 2020 and 2021 initiatives, and we have to yet see its full recovery.

In summary, we have not yet seen concerted, determined action (IPCC, 2022). And, of course, Covid-19 and a major war in Europe absorbed much of the public's attention. The impact of Covid-19 has been favourable in respect of emission reductions, with many flights grounded and lockdowns reducing local traffic. But little has been achieved on the emissions front and there is already a bounce-back to pre-pandemic levels of emissions, with governments and industry attempting to get everybody consuming again.

A ray of light comes from developments in renewable energy: it slowly replaces some fossil energy sources (Global Energy Review, 2021). This is unsurprising as there are commercial benefits to be gained. But progress is too slow to make a decisive dent in global emissions. Furthermore, the actions of householders are often triggered by government subsidies, rather than a commitment to contribute to reducing emissions. When these subsidies are reduced or eliminated, we see a reduction in the demand for solar panels, heat pumps, or electric vehicles.

In any case, there are no easy solutions left given that many decades have been wasted talking rather than doing. For all this time, a lack of action was tolerated as nobody felt an immediate risk to their wellbeing, the scientific community argued amongst itself, politicians did not want to rattle their constituents, vested interest in fossil fuel cast doubt over scientific findings, and economic progress was seen as more important and of immediate relevance compared to some future threat caused by global warming. And still, there are plenty of people who believe that some magic technological solution or scientific discovery will save us all…

Notes

1 For example, Cotgrove claimed that there is a stable and relatively enduring personality characteristic that predisposes part of the human race to believing in such technological fixes and the other part to believing that technology will destroy the planet. See Stephen Cotgrove, *Catastrophe or Cornucopia: The Environment, Politics, and the Future*, John Wiley & Sons, 1982.
2 We will cover dopamine, the 'feel-good neurotransmitter' in some detail in a later section.
3 *New Scientist* articles are available at Google books (books.google.com).
4 In the Hackett model presented earlier this relates to the second stage of the commitment to pro-environmental activities during which a person asks if the issue is of importance to them and to the third stage, when they ask about the effectiveness of any form of fix for the problem. Finally, in stage 4, they assess the extent to which they feel their own actions can make a significant effect on the issue.
5 There are numerous videos on www.bing.com covering the White House memo.

References

A Study of Climatological Research as it Pertains to Intelligence Problems, CIA (August 1974); the full text is available at www.governmentattic.com
BBC Radio Unveils 60 Years of Reith Lectures Archive, www.bbc.com, 26 June, 2011.
Bell, Alice (2021) *Our Biggest Experiment. An Epic History of the Climate Crisis*, New York: Counterpoint.
Global Energy Review (2021). Renewables, www.iea.org
Gordley-Smith, A., and Hackett, P.M.W. (2023) African Philosophy-Based Ecology-Centric Decolonised Design Thinking: A Declarative Mapping Sentence Exploration, *Filosofia Theoretica: Journal of African Philosophy, Culture and Religions*, V12(2). https://www.ajol.info/index.php/ft/article/view/260077
Hackett, P.M.W. (1995) *Conservation and the Consumer: Understanding Environmental Concern*, Abingdon: Routledge.
Hackett, P.M.W. (2014) *Facet Theory and the Mapping Sentence: Evolving Philosophy, Use and Application*, Basingstoke: Palgrave Macmillan Publishers.
Hackett, P.M.W. (ed.) (2018) *Mereologies, Ontologies and Facets: The Categorial Structure of Reality*, Lanham, MD: Lexington Books.

Hackett, P.M.W. (2019) *The Complexity of Bird Behaviour: A Facet Theory Approach*, Cham, CH: Springer.

Hackett, P.M.W. (2020) *Declarative Mapping Sentences in Qualitative Research: Theoretical Linguistic, and Applied Usages*, London: Routledge.

Hackett, P.M.W. (2021) *Facet Theory and the Mapping Sentence: Evolving Philosophy, Use and Declarative Applications* (second, revised and enlarged edition), Basingstoke: Palgrave Macmillan Publishers.

Hackett, P.M.W., and Lustig, K. (2021) *An Introduction to Using Mapping Sentences*, Basingstoke: Palgrave.

Jastrow, Robert, Nierenberg, William, et al. (January 1, 1990) *Global Warming: What Does the Science Tell Us?*, Arlington, Virginia, United States: George G Marshall Institute

Luntz, Frank (n.d.) For details see Frank Luntz on Wikipedia (en.wikipedia.org)

Lustig, K., and Hackett, P.M.W. (2020a) *The Philosophy of Facet Theory Pocket Guide*, San Francisco: Blurb Publishing.

Lustig, K., and Hackett, P.M.W. (2020b) *Mapping Sentence Pocket Guide*, San Francisco: Blurb Publishing.

Rizzo, Salvador, and Nirappi, Fenit (March 7, 2022) Global Covid-19 Death Toll Tops 6 Million, Another Grim Milestone in the Pandemic, *The Washington Post*. https://www.washingtonpost.com/health/2022/03/07/6-million-covid-deaths/

Schulz, Kathryn, Barron Mia, et al. (2010) *Being Wrong: Adventures in the Margin of Error*. Ecco, Jan 4, 2011.

Steidl, Peter (2017) *The Book of Change. Make the Changes You Want, and Make Them Stick*, Scotts Valley, California, United States: CreateSpace Independent Publishing.

Thatcher, Margaet (1989) *Margaret Thatcher Speech to the UN General Assembly (Global Environment)* www.margaretthatcher.org

The History of Earth Day (n.d.) www.earthday.com

The Story of Climate Change Right Now in 9 Charts, IPCC Report, last updated April 5, 2022.

UN Report (n.d.) *Report of the World Commission on Environment and Development: Our Common Future*, available at sustainable development.un.org

UNFCCC (n.d.) For details see Conference of the Parties (COP) at www.unfccc.int

United Nations Conference on the Human Environment, 5–16 June 1972, Stockholm, www.un.org

Wallace, David Foster (2006) *Infinite Jest*, New York, United States: Back Bay Books.

Why TIME Devoted an Entire Issue to Climate Change, 2050: The Fight for Earth (n.d.) www.time.com

Wikipedia (n.d.) An overview on Greta Thunberg's activism can be found on Wikipedia

2

WE CAN SUCCEED!

We need to ask ourselves a serious question: Have we become doom-mongers who see inaction and failure everywhere rather than progress and betterment? In our efforts to learn more about the challenges humankind faces, we have delved into the dispiriting world of failure. But there were also success stories. While they may not concern major global challenges, they highlighted that it is possible to solve problems that are widely seen as intractable. Is there anything these cases have in common? What can we learn from these isolated examples that may allow us to address major global challenges more effectively?

Over the following pages, we introduce you to six success stories from different times and geographies that prove that we can succeed against all odds.

United States, 1930

The Great Depression, the longest and deepest downturn in the history of the United States lasted more than a decade, from 1929 to 1941 (Digital Public Library of America, n.d.). In the 1930s another development affected people living in the mid-west and southern Great Plains. Droughts and land erosion due to unsustainable farming practices rendered some 35 million acres of formerly cultivated land useless for farming, while another 125 million acres was rapidly losing its topsoil (History, n.d.).

Solution

In 1933 the newly elected president, Franklin D. Roosevelt, initiated the creation of a Civilian Conservation Corps (Civilian Conservation Corps Legacy, n.d.). It was initially composed of only unmarried men aged between 18 and

DOI: 10.4324/9781003477167-2

23 years, but this restriction was eased in later years. Members of the Corps received $30 a month (equivalent to approximately $700 today), of which $25 was sent to their families, thereby spreading the funds across the United States.

In tune with its time, the Corps discriminated against women, who were not allowed to join, and African Americans, who were accommodated in separate camps. In fact, only 200,000 African Americans were allowed to join the programme, out of a total of just over 3 million participants. This is even more significant given that the Great Depression caused unemployment for 25% of white, but 50% of African American young men.

With respect to Native Americans, a separate stream called the 'Indian Division' was created which focused on carrying out conservation work in reservations. Tribal Councils participated in the administration of the works.

The members of the Corps benefitted from free meals and a place on one of the more than 4,500 camps that were established across the country. They were also permitted to take part in educational programmes after work.[1] This allowed some 57,000 men to learn to read, with some going on to gain a high school diploma or even a college degree. However, life was strictly regulated, with lights out at 10 p.m. and rising to get ready for work at 6 a.m.

The CCC renovated and created infrastructure, including new roads and trails, picnic areas, fireplaces, tourist cabins, ranger stations, flood control measures such as dams, and terraces to avoid further land erosion. They also worked on nature conservation and wildlife protection programmes, cultivated plants in nurseries and collected seeds for revegetation. They were also used as manpower in the response to acute disasters, fighting forest fires, snowstorms, and floods.

By the time the CCC initiative ended, the 3 million men on the scheme had planted 2.3 billion trees, saving 20 million acres of soil from erosion. They also created several national parks in the eastern United States, and completed the Appalachian trail, trails on Mount Mansfield, and the Colorado River trail in the Grand Canyon. Much of this infrastructure continues to serve the United States almost a century later.

Is the Concept Still Relevant Today?

It could be argued that there is a case to be made to launch a modern version of the CCC today.

First, the impact of climate change is becoming increasingly severe, and we need to move towards net zero emissions as fast as possible. Trees absorb CO_2 and this is stored when they are cut down. However, the stored carbon is released into the atmosphere when the wood is burned. This means that wildfires add significantly to our emission problem and the frequency and severity of wildfires has increased due to climate change, with these fires adding to emissions and thus contributing further to our crisis.

Second, youth unemployment is at nothing like the same rates it was during the Great Depression; furthermore, there are safety nets in many countries which ensure that the impact of temporary unemployment is not threatening to health or even life. Nevertheless, youth unemployment is at relatively high levels in several developed countries:

- the European Commission reports a youth unemployment rate of 14.2% in the EU and 14% in the euro area, with the incidence rate increasing slowly (European Commission, n.d.).
- In the US the youth unemployment rate was 8.7% in July 2023 (Bureau of Labor Statistics, n.d.).
- In China, the urban youth unemployment rate has risen to 21% as of May 2023 (CNBC, n.d.).

Finally, there are massive infrastructure challenges – in the US because of aging infrastructure which is deteriorating even faster in some regions due to extreme weather events. In China, meanwhile, infrastructure can hardly keep up with the massive pace of urban development. It may well be, therefore, that a modern version of the CCC could make a useful contribution to dealing with the devastation created by climate change while offering employment to young people.

Australia, 1949

The Refugee Crisis has been with us seemingly forever, with wars adding hundreds of thousands to their numbers and no strategies to address this issue. This is, of course, also the case with the war that is raging in Ukraine as we write. And recently we have seen Israel waging war against Palestine, killing so far some 40,000 civilians and causing a mass exodus from the affected area.[2]

The US and its allies have invaded countries, destroyed people's lives, and then left without giving any thought to restoring the damage done and putting millions on the path into refugee camps or worse. The EU is currently paying Turkey billions of dollars to keep refugees in camps away from the EU's borders. Australia spends hundreds of thousands for each refugee kept in prison-like overseas detention camps.

Is it possible to find a solution that provides refugees with a new life while also benefiting the country that takes them in? The following case study suggests that it is possible, at least in a localized way. Can we learn something from this example that could inform today's strategies? Let's have a look at how Australia changed its immigration policies to address a domestic problem and, in the process, helped tens of thousands of refugees to find a new life.

Today, Sydney's population is just over five million. But it would still be just a small city if not for a decision in 1949 that allowed Sydney to grow, revolutionized the economy, and put an end to Australia's discriminatory, UK-centric immigration policies.

Before the Second World War, Australia's economy was essentially built on agriculture with low electricity demands, but the war changed that. It led to a boost in industrial production which significantly increased demand well beyond existing generation capacity. The resulting severe power outages during the late 1940s and early 1950s put the government under pressure to find a solution. However, it was not just electricity that restricted Australia's development. The growing population led to increased demand for fruit, dairy, and other intensive agricultural products, which could not be satisfied without extensive irrigation.

Evaluation studies pointed to the Snowy Mountains in Southern New South Wales as the most appropriate area to develop a hydroelectric facility that could satisfy the demand for a substantial boost in both electricity generation and irrigation. To avoid politics diverting this project, the federal government declared the project a national security issue – Australia needed a reliable source of electricity away from coastal areas, which were vulnerable to attack. This allowed the federal government to introduce legislation under its defence powers, and thus the *Snowy Mountains Hydro-Electric Power Act 1949* came to pass (National Archives of Australia, n.d.). The Snowy Mountains Hydroelectric Authority (SMHEA) was established, and construction started on 17 October 1949. Prime Minister Chifley declared that Australia was *'on the threshold of a new era of great industrial and rural development'*.

While politicians may be prone to exaggerated claims, this was not one of them: The Snowy Mountains development was designed to spread across an area of 5,124 square kilometres (1,978 sq. miles), requiring the construction of seven power stations, 16 dams, 145 kilometres (90 miles) of tunnel, 80 kilometres (50 miles) of pipelines and aqueducts, 1600 kilometres (994 miles) of roads and railway tracks, seven townships, and 100 temporary camps.

The plan was sound and was a national priority but Australia, with a population of just eight million and a limited industrial workforce, lacked the labour force and expertise to realize it.

Solution

Two years after the end of the Second World War, some 850,000 people still lived in displaced persons camps across Europe, bringing together people from Armenia, Poland, Latvia, Lithuania, Estonia, Yugoslavia, Greece, Russia, Ukraine, Hungarian, Czechoslovakia, and other countries. Various countries invited Displaced People to work in industries where labour shortages were causing severe restrictions. For example, Belgium accepted 20,000 coal mine workers in 1947, and the United Kingdom opened its doors to 86,000 displaced persons as part of various labour import programmes.

At the time Australia's immigration policy was UK-centric. But the country's dire need for labour and expertise to address its electricity and irrigation deficiencies led to a broadening of its immigration restrictions in late 1947, changing the country's entire culture. The Snowy Mountains Authority brought in 182,159 Displaced People, mainly from Eastern Europe, under a two-year work scheme. In addition, about 600 highly skilled tradesmen and top surveyors were actively recruited in Germany.

Over the 25 years of the project, more than 100,000 men and women worked for the Snowy Mountains Hydro-electric Authority. Migrants from 32 (mainly European) nations made up around 65% of the scheme's workforce. These 'New Australians' provided skills and manpower. The government covered their travel costs, which they later repaid out of their earnings during their two years working on the Snowy Mountains scheme. Wages were set in accordance with the severity of the task and the associated risk: for example, working on the tunnels attracted wages up to six times the average.

Because of the increase in the number of immigrants, Australia established migrant reception centres, including three centres for non-English-speaking Displaced People from Europe and 20 holding centres to house non-working dependents. The purpose of these reception and training centres was to facilitate general medical examinations, issue necessary clothing, pay social service benefits, interview migrants to determine their employment potential, and provide instruction in English and the Australian way of life.

Result

This great influx of foreign skilled and unskilled workers engaged in building Australia's most significant energy infrastructure positively influenced national attitudes and government policy towards non-British immigration. It was still a 'White Australia' policy, but it did at least signal a first step towards greater diversity.

The Snowy Mountains Hydro-electric Scheme is one of the world's most complex integrated water and hydro-electric power schemes. It is listed as a "world-class civil engineering project" by the American Society of Civil Engineers. The Snowy Mountains Scheme continues to generate power today, producing 67% of all renewable energy in the mainland National Electricity Market, and providing water for an irrigated agriculture industry worth about AUD$3 bn per annum, and representing more than 40% of the gross value of Australia's agricultural production.

Notably, the scheme was a win–win development. Displaced People found a new home, had immediate employment, and became part of a diverse community. Meanwhile, Australia could address its electricity and irrigation challenges by creating a world-leading hydroelectric and irrigation system that changed Australia's economy and culture and allowed Sydney to grow into a major metropolis.

Austria, 1961

We are in Austria in the 1960s, observing what looks like a group of old friends having a good time. There is plenty of food and drink to enjoy while they animatedly share the latest jokes and stories. But they are not friends. By all accounts, they are enemies. These men are representatives of the Chambers of Commerce, the labour unions, and the Austrian Government, locked into wage negotiations that would affect millions of workers. In this case, 'locked' is used literally: they have been locked into a room and supplied with ample food and drink until they have reached an agreement. After all, sharing food and drink brings people together.

It is not easy to resolve an issue when it involves three parties, each with conflicting objectives and a need to get a positive outcome for their respective constituencies. These negotiations go on for untold hours. Eventually, however, an agreement is reached.

This may well be an embellished version of what really went on in these wage negotiations, as it comes from the father of one of the authors, who represented the Chambers of Commerce in negotiating wages and conditions for retail employees. He returned from the negotiations to tell the latest jokes he had picked up at these sessions, how they shared food and plenty of drink, and the shared understanding that each party needs to leave with something that benefits their constituency. In other words, the focus was on making sure everybody had won something of sufficient importance to their constituency, rather than pursuing a 'winner-takes-all' approach.

As none of the three authors has lived in Austria for many decades, we don't know if they still conduct wage negotiations like that. We hope they do because it worked and because it preserved the importance of each of the three parties and the dignity of the negotiators. But we somehow doubt that something as human as this approach would still be acceptable today.

Of course, you might think that this is a highly unprofessional way of reaching an agreement. But during the many years during which these types of negotiations took place, Austria measured labour strikes in minutes per year (Table 2.1) while other countries that pursued a more confrontational approach, such as France, the UK, and the US, lost many days – sometimes weeks – to labour unrest. An OECD report shows the following (OECDilibrary, n.d.).

Confrontational negotiations rarely work unless there is a dominant party who can bully the others into submission. Many challenges can only be addressed effectively when parties respect each other and are prepared to

TABLE 2.1 Annual average workdays lost per 1,000 salaried employees

Year	Austria	France	UK	US
1990	–	528	84	55
1995	–	784	19	51
2000	1	581	21	161
2005	–	164	9	13
2010	–	318	15	2
2015	–	n/a	6	5

make compromises. Further, bullying another party may give the bully a win today, but it creates resentment that will build up and lead to consequences down the track. Although, of course, today's bully may not care about these future consequences; they may be more than happy to leave it to their successors to deal with the fallout.

Collaboration Enshrined in Law

In Germany, public companies must establish two boards: a management board (Vorstand) comprised of executives and a supervisory board (Aufsichtsrat) made up of non-executive directors. The latter oversees and appoints the management board members and must approve major business decisions. Every company with more than 2000 employees must establish a Works Council representing the company's workers. The Works Council nominates half the supervisory board members under the country's two-tier structure. In companies with between 500 and 2,000 employees, the employees elect one-third of supervisory board members (worker-participation.eu, n.d.).

It may seem surprising that Germany gives proper representation to workers at a level where critical policy decisions that shape the company's future are being made. Interestingly, however, we also find a similar arrangement in China, another economic powerhouse (china.org.cn, n.d.). As in Germany, limited liability companies in China must have both a board of directors and a board of supervisors. The latter comprises shareholder and employee representatives. The Chinese corporation law specifies that the company's employees shall democratically elect the employees' representatives through a meeting of employees' representatives or a meeting of all employees.

The point we want to make here is that the second- and third-largest economies in the world take a collaborative approach to make corporate decisions. As much as executives and legislators in other countries claim that such an approach would cripple their country's economy, there is solid and irrefutable evidence to the contrary: collaboration beats confrontation.

South Africa, 1994

In the late 15th century, the Portuguese explorer Bartholomeu Dias became the first European to explore South Africa. In the 17teenth century, the Dutch took power in South Africa until the British colonized the country at the end of the 18th century. The British remained in control of South Africa until it became a republic in 1961.

For many years before the departure of the British, racial segregation, or apartheid, had been practiced and this continued for thirty more years after they departed. However, apartheid attracted increasing condemnation around the world. In 1974 they established a set of principles, known as the Mahlabatini Declaration of Faith, that called for equality for black and white people through a peaceful relocation of power. The move to a representative government in South Africa took several decades that were often character-ized by violence. However, the final stages of the collapse of white rule and the removal of apartheid happened in a few short years. The black leader, Nelson Mandela, had been held in prison for 27 years, but he was released in 1990 and the African National Congress (ANC) and other liberation groups were legitimized. In 1993, the white leader, F.W. de Klerk, and Nelson Mandela, began discussions about how such a transition could occur.

Democracy in South Africa was fought for, and had to be assembled, over a period of time. Towards the end of the era of apartheid, there was consider-able violence due to the urge for change and resistance to such change by some white South Africans. This violence, coupled with international eco-nomic and cultural sanctions, economic difficulties in the country, the end of the Cold War, along with fears that South Africa would enter into an internal racial war, all contributed to the collapse of white minority rule.

However, in what may be an unprecedented move, and something which has not happened since, the white South Africans essentially voluntarily relin-quished their hold on political power in the country. First, there was a referen-dum in 1992 in which the white voters approved government negotiations to end apartheid. Many thought that this would usher in a period of guerrilla war and civil war, but the final transition to a democratic form of government in South Africa has been called a 'miracle'. This name is due to the transfer being remarkably peaceful and the first universal elections taking place in 1994.

The ANC won the election and has remained in power up to the present. South Africa has changed in many ways since the end of apartheid. There have been, and still are, many challenges that face the country. However, in his Nobel lecture of 1993, F.W. de Klerk summed up what the removal of apartheid meant to his country and indeed the world:

What is taking place in South Africa is such a deed – a deed resounding over the earth – a deed of peace. It brings hope to all South Africans. It opens new horizons for Sub-Saharan Africa. It has the capacity to unlock the tremendous potential of our country and our region. The new era which is dawning in our country, beneath the great southern stars, will lift us out of the silent grief of our past and into a future in which there will be opportunity and space for joy and beauty – for real and lasting peace.

Iceland, 1997

Tell teenagers not to do something, and there is a fair chance they will find it even more exciting to do it. What can we do to address drug and alcohol use amongst teenagers? Obviously, the answer is not to wag a finger at them. But what is a more effective approach?

Surveys conducted in Iceland in 1992 and again in 1998 showed that the percentage of 15- to 16-year-olds who smoked cigarettes daily increased from 15% to 23% and the percentage of those who smoked cannabis rose from 7% to 17%. 40% of respondents stated they had been drunk in the past month, and 21% claimed to have drunk alcohol ten times or more in the past 12 months. Similarly, 14% claimed they had had an accident or injury related to alcohol (Charlie Sorrel, 2017). (Today, it is more likely that drugs would have been the preferred illicit substance to either escape from reality or get a dopamine hit, but we are talking about 1997.)

Solution

Iceland hired US psychology professor Harvey Milkman to help address this problem (Milkman, n.d.). Milkman's research in New York and Denver suggested that drugs and alcohol are people's ways of dealing with stress. He founded Project Self-Discovery in Denver, giving kids alternatives to drugs and crime. He believed that *"... drug education doesn't work because nobody pays attention to it"* and suggested that young people using drugs were addicted to a change in consciousness that allowed them to deal with stress, rather than to the drugs as such. Consequently, he gave them something better to do. Kids could choose to learn music, dance, hip hop, art, and martial arts. The programme also taught them life skills. It was a success.

Soon after his Denver success, Milkman was invited to Iceland to discuss his work and this led the establishment of a multifaceted programme called Youth in Iceland. He also made Iceland his new residence of choice.

The main features of the Youth in Iceland programme were: Alcohol sales were banned for anyone under 20, and the minimum age for buying tobacco was set at 18. Kids aged 13 to 16 were placed under a curfew: 10 p.m. during winter and midnight during the summer. Each local district came up with a communal pledge that parents could sign. These pledges specified commitments such as not allowing their teenage children to have unsupervised parties, that they would not buy alcohol for minors, or that they would take an interest in the wellbeing of other children.

However, the core of the concept was to teach new skills to young people aged 14 and over and give them positive experiences which would produce a natural 'high'. The programme was therefore built around offering incentives rather than imposing and enforcing restrictions. Families were provided with a Leisure Card which allowed each child to spend $300 a year on activities. More than 100 different types of organizations were associated with the Leisure Activities Card, including sporting clubs, dance schools, youth groups, and music schools. There was something for everyone.

This strategy is consistent with studies showing that children involved in organized recreational activities are less likely to engage in antisocial behaviour or become socially isolated. Instead, they will adapt more quickly to new communities and make new friends, boosting their self-esteem and self-image.

Result

In the programme's first 15 years – from 1997 to 2012 – teenagers spending time with parents doubled, as did the number engaged in organized sports. At the same time, cigarette smoking, drinking, and cannabis use in this age group all declined dramatically.

Today, Iceland tops the European table for the cleanest-living teens. The percentage of 15- and 16-year- olds who had been drunk in the previous month plummeted from 42% in 1998 to 5% in 2016, and the percentage who had ever used cannabis is down from 17% to 7%. Similarly, those smoking cigarettes on a daily basis fell from 23% to just 3%.

Iceland's model has been adopted by several cities, initially in Europe under the Youth in Europe programme and eventually more widely under the Planet Youth programme (ecad.net, n.d.). The surveys and programmes are tailored to fit local needs and problems. Alcohol and drug use are dropping in all of the participating cities.

Brazil, 2012

With tears in his eyes, Adriano dos Santos assures us, "I promise your eyes will keep on watching Sport Club Recife." "I promise that your heart will continue to beat strongly for Sport," says another person. Luiz Antonio promises that "Your lungs will keep on breathing for Sport Club Recife," while somebody else states, "Even a guy who backs another club, he will breathe Sport Recife."

Yes, these people are talking about Brazil's Sport Club Recife, but this story is not about soccer; it's about organ transplants.

It is extremely difficult to get people to opt-in when it comes to organ donation. However, in countries where you need to opt-out if you do not want to donate your organs, we find that few people bother to do so. This suggests that most people can't be bothered dealing with this issue. And why should they? They don't expect to suddenly find themselves in need of an organ, nor is there any reward for becoming an organ donor. You won't get any thanks from the person benefitting from your organs; you won't even be able to shake their hand, as you will be dead.

Under Brazilian law, the surviving family decides whether or not to donate the deceased family member's organs. But, of course, if the deceased has an organ donor card, they have decided for themselves. Therefore, it is vital to get people to opt-in and get an organ donor card.

Solution

Our story tells of events in Pernambuco, Brazil, where the local organ transplant centre could not help many patients that needed a transplant to save their life (Paul Smith, 2014). Or that was the case at least until 2012, when Sport Club Recife, one of Brazil's biggest domestic soccer clubs, kicked off an organ donation campaign. A video now plays before every home game in Sport Club Recife's 35,000-seat stadium, encouraging supporters to apply online for a Sport Donor card, a credit card-sized card featuring the outline of a heart with a fiery red backdrop. Many of the people in the video are Recife fans who were on transplant waiting lists when the campaign began (Julia Carneiro, 2014).

Sport Club Recife has some of the most passionate fans in Brazil. "1st God. 2nd Sport Club Recife. 3rd Family. 4th work," says one fan. "Nothing else matters. Sport Club Recife is everything!" says another. But they are passionate about life too – even after death. The club mobilized its devoted fans to save other people's lives, and 66,000 fans signed up because of the campaign.

Recognizing the influence it has over many thousands of soccer fans, the leadership of Sport Club Recife decided that they should use their power "for bigger things." Jorge Peixoto, the club's vice president for social programmes, believes that "every Brazilian is born with football in the soul." That belief forms the core of the team's movement to create a legion of "immortal fans" by having them donate their organs after their death.

Adding to the campaign's authenticity, the agency recruited actual patients waiting to receive transplants to help make the passionate appeal. Let's revisit the comments made by organ recipients and donors:

Adriano dos Santos, awaiting an eye transplant: "I promise your eyes will keep on watching Sport Club Recife." After receiving his new eyes: "My vision is returning. I feel like I was born again."

"I promise that your heart will continue to beat strongly for Sport," says one of the patients. Luiz Antonio, waiting for a lung transplant: "Your lungs will keep on breathing for Sport Club Recife," and a donor says, "Even if I donate my lungs to a guy who backs another club, he will breathe Sport Recife."

A woman who received a heart transplant: "My new heart comes from a Sport Club Recife fan, and it will keep on beating for Sport Club Recife."

The "Immortal Fans" initiative has been a success. More than 66,000 soccer fans have signed up for the card. In the campaign's first year, the transplant waiting list was reduced to zero in Recife, while the number of heart transplants in the region increased more than fourfold. The sheer number of donors in Recife has led to more transplants throughout the surrounding Brazilian state of Pernambuco (Angelico Law, 2014).

Germany, 2018

'No one gets left behind.' An age-old pledge in military circles, adopted by a few aid organizations like the Refugee Council.

The promise to leave no one behind is powerful, especially when the future is uncertain. But when we are dealing with large-scale social issues, it requires a good deal of determination to put it into practice. What happens when a community is confronted with major economic change due to its established employment base disappearing? This is the challenge Germany faced when embarking on the government-mandated closure of coal mines.

A Tale of Two Countries

Coal is essential to addressing Germany's power requirements for its 84 million people and the world's third-largest economy. The country is also the world's largest lignite producer. Therefore, it may come as a surprise that Germany shut down its last black coal mine in 2018 and that it is currently executing a programme to close lignite or brown coal mines (Eric Campbell, 2020; Nick O'Malley, 2019).

The German government and regional leaders agreed to phase out coal-fired power stations by 2035. But they are not using strongman tactics to achieve this: lignite mine and coal-fired power plant workers will benefit from an investment of $45 billion. A significant share of this money will fund new infrastructure projects in the affected areas, offering new employment opportunities. The objective is to leave no worker behind, i.e., to provide new jobs for everybody who has lost employment because of the closures and wants to remain part of the workforce. But even shareholders and investors are being looked after. Coal-fired power stations will receive compensation of somewhere between €1.75bn and €2.6bn.

It is also worth noting that Germany is working towards phasing out nuclear power by 2022 as its response to the Fukushima nuclear disaster. Rather than relying on coal or nuclear energy, Germany aims to generate at least 65% of its electricity from renewables by 2030.[3]

Contrast Germany's approach with what is happening at the same time in Australia, where new coal mining licenses are being issued. For example, a recent permit to establish the massive Adani Carmichael mine will generate significant coal output. While Germany is cleaning up its act, Australia is adding more to the world's emissions.

We need to consider the difference in the approach taken to appreciate how an enlightened country can deal effectively with complex challenges such as phasing out a well-established energy source. In contrast, in a country led by short term-thinking, ignorant leaders not only fails to take corrective action but adds even more to emissions. In Germany, trade unions, energy companies, green groups, and the government worked to find a common cause. All parties agreed that the coal industry must be closed. Good-faith negotiations could then proceed, ensuring that the people these parties represent would not suffer unfairly. Naturally a major restructuring of the industrial landscape takes time, but progress is being made and critical milestones are being reached.

The process has not been without the occasional demonstration or negative publicity; for some, the process of closure may take too long, while for others it may be too fast. But the programme is succeeding, and Germany will make a significant contribution to fighting climate change at a global level.

Australia could have adopted the 'German approach', developing infrastructure projects and compensating workers and shareholders while going through a well-considered programme of shutting down mines and coal-fired power stations. But this was not to be. In 2017, Scott Morrison, then the Liberal Government's Treasurer, made clear his lack of commitment to fighting climate change when he brought a lump of coal into parliament and proudly declared his support for coal mining. When taking over as prime minister, his position did not change.

In 2019, Australia's coal mines employed 52,600 people (according to the Labour Force Survey, February 2019). That's not a large number, given Australia's population of 26 million. But there are two barriers to taking an enlightened approach in this area. First, employers, trade unions, and the government do not work closely together. Second, the coal mines are in marginal electorates, and politicians are typically more interested in votes and winning elections than building a better future for their country. It therefore came as no surprise that the Labour Party, the only significant alternative to the governing Liberals, also endorsed coal mining and did not object to issuing new mining permits.

If we want to achieve wholesale change we need to make sure we don't leave anyone behind. Fairness is essential, and a belief that fairness will prevail provides security and a sense of belonging. Everyone's focus can therefore be on how to achieve a shared objective and create an alternative future, rather than personal interest. But we also need brave and intelligent leaders with sound values to lead the change. Germany ticked all the boxes here, Australia not even one.

Conclusions

The US president established a Civil Task Force that helped revitalize the environment in the mid-west and Southern Great Plains which had been destroyed by human ignorance and greed; Australians took in displaced people; Icelanders helped young people to escape a world of alcohol and smoking; South Africa abolished apartheid when two enlightened leaders were able to be face-to-face;[4] Brazilians banded together to boost organ donations; Austrians put respect and collaboration ahead of pigheadedness and political point-scoring; Germans were, and still are, united in closing coal mines to eliminate a major climate change driver.

But – and here comes a big 'but' – what have the other 7.8 billion people done? Why don't we see these programmes, adapted for local conditions, implemented around the world? Why do we have 110 million displaced people, including 36.4 million refugees with many living in camps in substandard conditions? Or, even worse, why do Australians put refugees in offshore prison-like detention centres and deny them even the most basic care? Why do we see conflicts between political parties, between business and labour unions, between rich and poor, between different races and religions, rather than a united effort to address the challenges we all face? Why do we see people suppressed and even killed for their religious affiliations or the values they hold? Why do wealthy nations invade other countries, destroy their infrastructure and people's homes, and then leave without shame, ignoring the devastation they leave behind? Why do governments, interest groups, and individuals knowingly spread fake news? Why do we destroy the environment? Why do we keep talking about the climate crisis but do far too little to address it?

In Chapter 1, we saw the same patterns of behaviour regardless of the global challenge. There is clearly a Commitment Gap. How has this Commitment Gap been closed in the success stories we have just reviewed? In all these cases, the solutions were aligned with the way our mind works or, if you prefer, with human nature.

The US president implemented a bold approach to boost employment during the Great Depression, while also restoring a devastated environment that had been lost to agriculture. It is arguably not difficult to get the public's support for an employment creation scheme during a severe Depression, nor is it unexpected to get the support of those who benefitted directly from this initiative. However, it is noteworthy that, rather than resorting to subsidies that would have been just a short-term band-aid solution, the action taken addressed the challenge head-on.

In Australia, the need for increased electricity output to fuel growth was matched with the need to find new homes for displaced people, leading not just to a successful major infrastructure project but lasting changes to the country's immigration policy and culture. The decision to broaden immigration restrictions were aligned with the goals of Sydneysiders: to have secure electricity supply.

Austrians learned, from the war and years of occupation, that working together is more effective than competing and benefitted from the results of a non-confrontational culture. Iceland, meanwhile, accepted that teenagers would look for a dopamine hit, and that the only way to shape their behaviour was to offer them desirable activities that delivered one. The Recife Sport Club case shows that we can get action when we shift the goal from one that is not relevant to one that is. Finally, Germany demonstrated how 'belonging' could work, finding support for coal mine closures even among those most affected when they promised that nobody would be left behind.

We need to recognize that it is natural for us humans to show a lack of commitment when facing major global challenges that do not yet present an imminent and personal threat. Nor will we actively support solutions that do not benefit us. Until we accept this, we will continue to waste time with inadequate measures, arguments, and discussions. But what exactly is human nature, and why is it putting us on the path to destruction? And, finally, how is it that we have seen unexpected successes that seem to deviate from what we consider our ingrained approach to dealing with challenges?

Notes

1 Note that the camps move on once their work had been completed in a particular location. In other words, while hundreds of camps existed concurrently, 4,500 camps was the total number over a ten-year period.
2 While we totally support Israel's right to defend itself and find Hamas' attack on Israel an atrocity, we do not support the killing of civilians in the Gaza Strip.
3 We note that the energy shortages due to the reduced import of Russian gas as a reaction to its invasion of Ukraine may be slowing down the phase-out timelines. But the commitment to phase out coal-fired and nuclear power plants remains.
4 It should be remembered that there were also considerable international pressures upon the white government in Pretoria, and also fear of an internal race-based war.

References

Bureau of Labor Statistics (n.d.) *Employment and Unemployment Among Youth – Summer 2023*, https://www.bls.gov/news.release/pdf/youth.pdf

Campbell, Eric Germany Is Shutting Down Its Coal Industry for Good, So Far Without Sacking a Single Worker, February 18, 2020, Foreign Correspondent. https://www.youtube.com/watch?v=b11RUFvLZcc

Carneiro, Julia How Thousands of Football Fans are Helping to Save Lives, *BBC Brazil*, June 1, 2014. https://www.bbc.co.uk/news/magazine-27632527

Civilian Conservation Corps Legacy (n.d.) http://www.ccclegacy.org/

CNBC (n.d.) Why Youth Unemployment is Surging in China, https://www.cnbc.com/2023/09/03/chinas-urban-youth-unemploymentcrisis.html#:~:text=The%20world's%20second%2Dmost%20populous,from%2015.4%25%20two%20years%20earlier

Companies Law of the People's Republic of China (n.d.) www.china.org.cn

Digital Public Library of America (n.d.) https://dp.la/exhibitions/civilian-conservation-corps/history-ccc

European Commission (n.d.) Unemployment Statistics, https://ec.europa.eu

History (n.d.) https://www.history.com/topics/great-depression/dust-bowl

Law, Angelico Organ Donations Campaign Saves Lives in Brazil, July 22, 2014. https://en.paperblog.com/organ-donation-campaign-saves-lives-in-brazil-956220/

Milkman, Harvey (n.d.) www.us.sagepub.com/en-us/nam/author/harvey-b-milkman. This contains details of Milkman's extensive writings on addiction and behavioural change.

National Archives of Australia (n.d.) Details can be found at: *Snowy Mountains Hydro-Electric Scheme*, www.naa.gov.au

O'Malley, Nick How Germany Closed Its Coal Industry Without Sacking a Single Miner, *The Sydney Morning Herald*, July 14, 2019. https://www.smh.com.au/environment/climate-change/how-germany-closed-its-coal-industry-without-sacking-a-single-miner-20190711-p526ez.html

OECDilibrary (n.d.) www.oecd-ilibrary.org. Note that 2020 is an aberration because of the impact of Covid-19 and has therefore not been included.

Smith, Paul *Immortal Fans, Sport Club Recife Donate Organs*, May 11, 2014, www.greatmomentsofsportsmanship.com

Sorrel, Charlie Iceland Fixed Its Teen Substance-Abuse Problem By Giving Them Something Better to Do, Fast Company, February 14, 2017. https://www.fastcompany.com/3067732/iceland-fixed-its-teen-substance-abuse-problem-by-giving-them-something-better-to-do

Workplace Representation (n.d.) www.worker-participation.eu

Youth in Europe – A Drug Prevention Programme (n.d.) www.ecad.net

3

HUMAN NATURE

Ignore It at Your Peril!

Think about public service campaigns to which you may have been exposed. They may have dealt with a wide range of issues: vaccination resistance, drink driving, using seatbelts, adhering to speed limits, organ donation, respecting the elderly, domestic violence, recycling, reducing the use of plastics or other environmentally conscious actions. There is every chance these campaigns had something in common: They tell you what you should do (or not do) and stress the consequences of failing to 'do the right thing'.

These public service campaigns rest on two underlying assumptions: First, they assume we want to do the right thing but are ignorant and need someone to tell us what that is. Second, that we are easily intimidated and, hearing about the possible consequences of our undesirable behaviour, we will be immediately, magically, reformed.

Large-scale behavioural change is possible. For example, we know that random breath and drug testing has significantly reduced drink driving, and that a dramatic surge in Covid-19 infections boosts the number of people wanting to be vaccinated. But the success stories are invariably aligned with how the human mind works and what truly drives our behaviour.

Road safety offers an excellent example of how important it is to align behaviour change programmes with the way our mind works. Research studies have repeatedly shown that the vast majority of drivers are convinced that their driving skills are above average. When exposed to a campaign suggesting they should keep to the speed limit or avoid driving under the influence of alcohol or drugs, these drivers will not see the message as relevant to them, because they believe themselves to have outstanding driving skills and therefore able to manage driving fast or when intoxicated. However, even the best driving skills will not save them from random breath tests, which is why

DOI: 10.4324/9781003477167-3

random breath testing is invariably more effective than a campaign focusing on the risks and consequences of driving under the influence, however horrific those consequences might be.

Unfortunately, aligning behaviour change programmes with the reality of what drives people's actions seems to be a low priority for both advertising agencies and their clients as we persistently see this ineffective approach taken again and again.

Einstein has allegedly been said to have remarked that insanity is doing the same thing repeatedly and expecting different results. So why does this ineffective use of taxpayer money to shape the public's behaviour persist? Our only explanation is that those involved are working to outdated behavioural models and have not caught up with the many insights into how the human mind works that have been delivered by a range of disciplines over the last few decades. Let us present the most important of these insights and explore how building our approach on this scientific foundation could lead to a much greater impact when trying to shape behaviour to address our major global challenges.

The Danger of Relying on False Assumptions

Does it matter whether or not something is intrinsically good?

Public service campaigns typically assume that people are 'good'. They want to do the 'right thing' and therefore it is sufficient to let them know what the right thing is. Example might include: don't drink drive because you might kill someone; become an organ donor to save the lives of others; pay your taxes to help fund social and health services for people who are less fortunate; get vaccinated to protect not just yourself, but the people around you; stop smoking and eat healthy food to improve your life and reduce the burden on the healthcare system; respect women and your elders; and on it goes.

Of course, if people wanted to do the right thing, but were unsure what that would be, these campaigns would be hugely successful. 'Oh!', men would say, 'I didn't realize we should respect women! But now that I know, I will certainly change my behaviour!' This sounds like a stupid example but think about it: Why don't campaigns of this nature trigger massive behavioural change? Why do they fail?

Given that government agencies spend vast sums of money, they tend to justify their campaigns with survey results showing that a large percentage of respondents *thought* about doing the right thing or *plan* to do the right thing after being exposed to a campaign ad. Of course, often behavior can be tracked, and the results show that while respondents may have stated the intention to do the right thing, they have not acted in line with their stated intent. That is no surprise: how many people say they plan to lose weight, stop smoking, work harder, spend more time with their family, or eat healthier food, yet don't act accordingly even though these personal goals are presumably important to them. What people say they will do and what they end up doing are often not the same.

The assumption that we all want to do the right thing and only need to be told what that is, leads to a waste of substantial funds that are invested into public service campaigns. So, what keeps the decision-makers from searching for fresh approaches that have a much better chance of working?

The sad truth is that most of us want to believe that humans are 'good by nature' and there is much profit to be made from telling people what they want to hear. Numerous programmes, apps, workshops, conferences, blogs, and books tell people that they are naturally 'good'. Self-appointed gurus promise to help you find your 'inner good'. Organizations and individuals have made hundreds of millions of dollars because their message aligns with how many people like to think of themselves, despite the lack of evidence to support such claims.

Sometimes the most elaborate contortions are used to prove that humankind is naturally good. One of us recently started reading a book in which the author proposed that we are good by nature, but that we don't always act in a positive way because we don't *believe* that we are good (Rutger Bregman, 2021). All we need to do is believe we are good and, bingo, the world is a better place!

Bregman's book starts with an extensive study on the bombing of London, followed by the bombing of a multitude of German cities towards the end of the Second World War. The author seriously suggests that these events prove that we are good by nature! How? Because, by and large, people in the affected cities did not trample others trying to escape, did not steal, rape, or murder, but instead helped each other. Wow! What about the people who ordered the bombing, dropped the bombs, or the many citizens who supported bombing their enemies into oblivion? How can the bombing of cities causing the death of untold civilians ever be proof that humankind is good? And, by the way, it is well established that having a common enemy – and somebody dropping bombs on your city would qualify here – does bring people closer. That's neither new nor proof that people are naturally 'good'. Such behavior is simply due to our chances of survival when under serious threat improving when we act as a group. In this context we note that looting is not uncommon when a state of emergency is called. In this case, there is no collective threat, such as an armada of bombers, to bond the community.

Of course, those responsible for major bombing raids on the civilian population often explain that this was necessary to achieve a decisive victory that will end the war, thus saving millions from more suffering and death. This was also, of course, one of the main arguments advanced to explain the use of nuclear weapons in Hiroshima and Nagasaki as well as the massive Allied bombing raids on German cities. At the time of writing, Israel has killed more than 40,000 civilians with bombings in the Gaza strip claiming this is necessary to ensure the country's safety. We are in no position to judge the claimed trade-offs, but we are absolutely certain that killing civilians is not proof that humans are good by nature.

If you want proof that humankind is not good by nature, just think about what happened when the Covid-19 pandemic engulfed the world (Europol, 2021; Louis De Gabriele and Abela, 2020; Martin Burnt, 2020): people selling fake cures to make a quick buck; a black market for forged vaccination passports emerging; people illegally jumping the vaccination queues without regard to those who needed more urgent protection; breaking lockdown rules and spreading the virus by doing so; verbally and physically attacking others who wear masks or line up to get vaccinated or who try to keep order such as airline flight crews and store assistants; ripping off government support schemes to make money. And don't get us started on politicians who – in many countries – put their ego ahead of saving people's lives. We can't resist inserting another one of the many quotes attributed to Einstein: *The difference between stupidity and genius is that genius has its limit.*

The question of human beings being either 'good' or 'bad' is complicated because we are talking about a value judgement. Consider the Taliban. The United States trained them and provided them with weapons to kill insurgent Russian soldiers. They were rewarded for doing so. At this point they might be considered 'good'. But when the Russians withdrew from Afghanistan and the Americans became the insurgents, the Taliban were suddenly 'bad' when they carried on doing exactly what they had done previously: killing insurgents.

Or take the issue of abortion: if you believe that a woman who has become pregnant through rape should be allowed an abortion, you will be seen as 'bad' by the Right for Life supporters, but you will equally see the Right for Life supporters as 'bad' if you believe a woman should be allowed to make that choice. The point is that 'good' and 'bad' are value judgements that may depend upon your political leanings, religious beliefs, personal situation, upbringing, and past experiences. A value system is, by nature, not objective. It is biased. We cannot hope to address today's problems effectively when we make decisions based on a subjective value system.

A value is a psychological construct, a 'something' such as a belief, that I hold to be important. It is not something that I have a vaguely positive orientation towards, but it is something that I invest mental time and energy into. It also tends to be quite broad and covers wide aspects of life. A belief in freedom of speech, for example, is not concerned with just a single incident of speaking out in a given situation, although this specific instance will be affected by this value. For people who hold this value, it applies to – and has an influence on – many or all situations where a person may speak out.

Importantly, value systems are complex, comprising of a network of values rather than a single value. This may complicate things because values may conflict. One example may make this clearer. If I am a person who holds right-wing values, I may support the dismantling of the welfare state and encouraging people to stand on their own two feet without a 'nanny state'. I may also support strict policing and tough penal sentences. However, on election day I may not vote for the candidate who puts forward these policies. The reason for my apparently contradictory behaviour being that the candidate was a woman and I think it important that the leader of the country should be a man. Human beings strive for consonance between their values, but often this is not possible and predicting the behaviour of others may be complicated by dissonance amongst values.

If we want to change people's actions, we need to develop interventions that rest on a sound understanding of what is driving their behaviour. Even if we don't like the idea, we need to accept that people don't always want to do the right thing. Humankind did not get to where we are today by being friendly and thoughtful. We got here by outsmarting and killing whatever was in our way, including our fellow humans. And while we may have moderated our behaviour to some degree, we are still driven by what got us to where we are today.

To make progress we need to stop judging human behaviour and focus on understanding it. Only if we understand the drivers of behaviour and align our interventions with these drivers will we have any hope of success. But, of course, we can't escape a value judgement at this stage either: How do we want to shape people's behaviour? What is the outcome we would like to see? We propose a simple objective: *We should strive to improve the lives of people who are worse off than us*. As you will see, this leads to a hierarchy of responsibilities: Everyone needs to contribute in some way, but some more so than others.

Drivers of Behaviour

Having decided what we should strive for, we need to now focus on understanding what drives human behaviour. While it may seem unlikely, the key to this understanding lies in the distant past.

Our brain developed over a period of some 4 to 5 million years and, over that period, refined its performance significantly. It is designed to help us survive in a hostile natural environment, and it still carries out this function even though most of the world's population today finds itself in a far more secure human-made environment. The human brain allows us to address fundamental survival challenges better than other species because:

- it provides us with hardwired responses that are essential to survival in a hostile environment, such as the 'fight or flight' response
- it drives us to actions that are necessary for our personal survival and the survival of our species and is highly effective in gathering information that can be accessed in the future to help us learn from the past, thus boosting our chances of survival.

But how did we get there? Some 80 to 100,000 years ago, something amazing happened: humans developed a 'new' brain (esciencenews, 2011). This makes the brain a unique part of our body: Over millions of years, the shape of the human body evolved, allowing us to perform specific tasks better. But the brain *did not change*. Instead, it added a whole new brain to allow us to deal with new, evolving challenges.

As humankind moved from a natural environment into a human-made environment, it became important for people to gain a better understanding of how things work, to understand the consequences of their actions, to weigh up the relative benefits of following one path or another, to plan, and to understand the complexities of societies and communities. These demands were met by developing this new brain, the pre-frontal cortex, which allows us to evaluate actions we are taking or are just about to take. It may restrain us from acting if it identifies severe negative consequences. Further, this new brain also gives us a conscious understanding of who we are, what we do, and where we fit in our world. Due to these developments, we ended up with two brains that complement each other. This sounds like a great deal, doesn't it?

It's certainly good that we have this wide range of capabilities. The 'old' brain drives what we feel and do, while the 'new' brain allows us to consciously think, plan, analyse and, importantly, moderate actions driven by the old brain. But there is a problem: These two systems work in parallel rather than in tandem, each carrying out their respective functions without necessarily being aware of what is going on in the other brain.

It follows that these two brains may even pursue conflicting goals. When this happens, which brain is likely to win? The old brain, primitive though it may seem, is fast and powerful. It manages the hundreds of millions of impressions your sensory organs transmit to your brain every day, deciding which to place into memory and which to discard. It is also responsible for memory retrieval and manages all your organs simultaneously, monitoring them and releasing hormones and other chemicals to ensure they function well, taking corrective action whenever required. Meanwhile, studies have shown that the new brain, our conscious mind, cannot even pursue two activities in parallel unless we have habitualized at least one of them. In fact, the old brain's nonconscious mind can process 11,000,000 bits of information per second, while the new brain that relies on conscious thought can process only 40 bits of information per second, making it look slow and feeble in comparison.

Unsurprisingly, therefore, most of what we do is driven by our old brain, even though you may be able to rationalize after the fact and come up with seemingly sensible reasons for your actions.

There are two key reasons why the old brain shapes most of what we feel, think, and do: First, some of the responses that helped humankind survive have been hardwired into our brains and are still driving our behaviour to this day. Second, some of the complex challenges we face today require significant cognitive processing. It takes effort to 'think things through' but, rather than wait for the slower conscious mind to work out what to do, the old brain chooses a shortcut solution, driving us to act before we have thought through the situation in full.

The dominance of the 'old' brain has some important implications for our behaviour.

Of 'Clear and Present Danger': Immediacy and Relevance

When you live in a hostile natural environment but don't have a brain that can analyse, project, and think through actions, you need to be highly selective in order to survive. If you give your limited attention to something that is not relevant because it does not affect you right now, you may miss immediate, personal threats or opportunities.

The ability to focus on – and react to – anything that is immediate and relevant increased the chances of survival and, by doing so, eventually developed into a hardwired brain circuit. Even when humans traded the natural hostile environment for communities that were well removed from nature, they carried this hardwired circuit with them.

This explains why people may agree that it is 'a good thing' to donate their organs, but take no action to do so. After all, they don't feel their demise is imminent or that they will need an organ transplant themselves – there is neither immediacy nor relevance to drive their actions. The Sport Club Recife initiative took the emphasis from organ donations by focusing on a more immediate and personal opportunity: A new way to show support for your beloved soccer club. The desire to belong is, again, a hardwired brain circuit, as those who belonged to a group were more likely to survive in a hostile natural environment.

You may have come across the term 'clear and present danger.' In United States law, this doctrine is used to test whether limitations may be placed on the right to free speech (as well as being the title of a novel by Tom Clancy, later turned into a movie starring Harrison Ford). In the context of our discussion, the critical point is that the concept of 'clear and present danger' is hardwired into the human brain.

We are designed to act when facing a clear and present danger – one that threatens us directly and immediately – or when we are presented with a clear and present opportunity. Behavioural psychologists talk about this as a stimulus-response or stimulus and reaction (S-R) link. We can make linkages between a stimulus and responses that are close together in time. This makes sense as our sub-conscious system is there to protect – for example, to just run as fast as we can when faced with danger. However, this natural response has the effect of making us less concerned about dangers with longer term consequences.

You may think of climate change, the technological revolution or food security as massive challenges we should all focus on and address as a global community. But if the situation affects you neither personally nor immediately, your brain is programmed to ignore it or rationalize it away. The issue may well become acute and personal one day, but until then you are programmed to discard it. And, while this is of course a highly effective way to operate in a natural hostile environment where you need to focus on the here and now to survive, it is not particularly helpful to human beings today.

The problem of our short-sightedness is compounded by the neo-Liberal capitalist economy and political system through its emphasis upon short-term personal gain. The free-market needs consumers to buy its products and use its services and under this economic system the individual's freedom to consume as they want is emphasized – to misquote Descartes: "I consume therefore I am". Unfortunately, whilst this boosts our personal happiness, which is ostensibly a good thing, the beneficial effects are only temporary. Happiness in the here and now is emphasized because, as we have already noted, we are programmed to react to immediate rather than longer-term consequences of our action. This has the unfortunate consequence of encouraging many forms of behaviour that have unwanted or undesirable effects in the longer term, such as the ecological effects of our consumption-based societies.

We are Boiling in a Pot of Familiarity

Another built-in process that limits our willingness to address challenges that develop incrementally over a longer period is our in-built ability to adapt. Again, this was an important quality which allowed humankind to adapt well to changing environments, may they be of a natural or a man-made kind. The problem today is that adapting to a developing challenge kills the actions that need to be taken to address it.

You have undoubtedly heard the story of the frog put in a pot of cold water over a low flame. As the water slowly heats up, the frog adapts to the incremental change in temperature that will eventually lead to his demise and so boils to death rather than jumps out.

We are all frogs. We can get used to even the most disturbing developments and events if they evolve at an incremental rate. The problem is that we start to feel a sense of familiarity when we are exposed repeatedly to the same warnings, experiences, and events. In a hostile natural environment, familiarity was hugely important as it suggested a safe environment or a well-tested way of dealing with a threat. Today, we are still hardwired to mistake familiarity for safety. Families grow up in the shadow of active volcanos. Villages and cities destroyed by earthquakes or devastated by wildfires get rebuilt despite the likelihood of another catastrophic event.

At the same time, we find that a lack of familiarity can create significant stress and make us feel insecure. One extreme example is prisoners who are released after serving long sentences and who re-offend specifically to be returned to the familiarity of a cell. We all tend to feel insecure when we don't engage with a particular activity for a long time. Many of us will have experienced anxiety around socializing after spending lengthy periods in enforced lockdowns or voluntary isolation during the Covid-19 pandemic. When an activity or situation no longer feels familiar, we start to feel inadequate, try to avoid the situation, and, by doing so, further fuel our feeling of inadequacy.

Familiarity is a key reason for adapting to, and accepting, negative changes to our world. To name just one example: According to surveys, most people do not respect politicians and expect them to care more about their next election than the citizens they are supposed to represent. But we have got used to scandal after scandal, ignorance, lack of attention to detail, subterfuge, pork-barrelling, and dishonesty. Today, no one seems terribly upset when politicians behave that way because our familiarity creates an expectation that is met whenever we see a politician exhibit these qualities.[1]

This was evidenced in the UK with the then prime minister, Boris Johnson, being charged by the police for his role in what has come to be called 'partygate' – allegations that number 10 hosted, and that politicians from the governing Conservative Party were involved in lockdown-breaking celebrations (William Booth and Adam, 2022). Even as the country was largely observing lockdown regulations, the PM has been accused of flaunting these. The point here is not whether or not he is guilty of committing such behaviours, but rather that one of us has experienced informed and educated friends commenting: 'As if it matters! Of all the crimes and bad things politicians do, breaking the lockdown rules to have a party isn't that bad.' We have come to accept that politicians lie, and we now ask if the lie bad enough to justify reprimand rather than if we can trust politicians that have been shown up to lie to us.

Similarly, we don't worry too much about fake news of the kind spread by ex-President Donald Trump about the 2020 election being stolen from him or the claims of Boris Johnson when campaigning for Brexit in 2016. We do not worry too much about our privacy being intruded on by search engines, software, and content providers, about incompetent public sector organizations, or about the paid lobbyist who shapes policies for his paymasters by influencing political decisions. This is just how it is, we think, and we are not upset by it because we got used to it.

Think about climate change: Do you expect a breakthrough at the next COP conference leading to concerted, collaborative action, after more than two dozen conferences have not delivered this? We are now used to the lack of progress, and there is no outrage except on the part of highly committed NGOs or demonstrators, but they represent a tiny part of the broader community.

The Role of Neurotransmitters

We have talked about how our brain is hardwired to deal in a certain way with our experiences, information, and observations. But what is activating these brain circuits? This question leads us into the world of neurotransmitters. We are not planning to deliver an expert review on this topic, but if we want to successfully change human behaviour it is important to understand the ingenious way our brain shapes what we feel, think, and do.

Receptors are located between neurons to facilitate the exchange of information, a process known as neurotransmission. These receptors can release a neurotransmitter or receive the signal the neurotransmitter sends. In summary, neurotransmitters are the chemical messengers the brain produces and sends to a target cell, which can be a nerve cell, a muscle cell, or a gland. This is how the brain regulates moods, feelings, the sleep cycle, hunger, pain, and much more, shaping how we experience sensory inputs and how we react to them. There are more than 100 neurotransmitters, but we will limit our discussion to just a few that trend to have the most significant impact on behaviour.[2]

The Hunt for Dopamine Drives Much of What We Do

What would you do to boost somebody's chances of survival in a hostile natural environment? What about the carrot-and-stick approach? But hold on, the stick would not work because the person would most likely be dead if they didn't do the right thing. But the carrot may work: What about delivering a reward every time a person does something right?

This is essentially how our old brain drives us to act. Whenever we take a positive action that may aid our survival, the brain releases dopamine, also known as the 'feel good neurotransmitter'. Dopamine plays a crucial role in the motivational component of the reward system, where it modulates desire and motivation. It also modulates attention, cognitive processing speed, working memory and positive affect.

We experience a dopamine hit when we have a positive experience, take positive action, solve a problem, have a burden lifted from their shoulders, are recognised, or rewarded, make others feel better, and so forth. We can get our dopamine hits by being liked or followed on social media, competing, playing online games or a sport, cheering on an athlete or sporting team, indulging in entertainment or holidays, or impressing others (either in real life or online). We are driven by our desire to enjoy the dopamine opportunities our life offers us, and we are busy – so very busy – chasing them (Anne Trafton, 2020).

Importantly, dopamine is also released when we expect something positive to happen. You may have experienced a situation where you look forward to meeting a friend, going to a restaurant or bar, expecting to find a 'like' or positive comment on social media, or generally expecting an enjoyable experience.

And when you imagine this, you will feel good – because your brain releases dopamine, just as it would if these events had already taken place. Indeed, the dopamine release triggered by expectations is typically stronger than the hit we get when we actually experience the positive situation.

But here is the clever part: While dopamine does indeed make us feel great, dopamine levels subside quite quickly, leaving us craving another hit. This drives us to repeat actions that have rewarded us with a dopamine hit in the past or to try something new that promises to deliver one. However, as we get used to more frequent and more substantial dopamine hits, we want more and stronger again, so that the hunt for dopamine drives us to act, move forward, do better.

In other words, while dopamine drives us to seek happiness (i.e., to look for another dopamine hit), it does not allow us to be happy. That is the whole point of how dopamine works: If we were truly happy, we would stop striving for more and, while this may avoid some of the problems we have created in today's world, it would almost certainly have led to our extinction in a hostile natural environment.

This is the very basic, yet most effective, cycle that has driven humans to develop better shelter, weapons, hunting techniques and, later, to explore new worlds, create new societal structures, advance commerce, create new entertainment and leisure options, and come up with new ways to compete that promise winners a dopamine hit. There is no doubt that dopamine is driving the advancement of technologies and their use, for better or worse. Dopamine is the neurotransmitter that has driven humans to dominate the world and it may well lead to the future demise of the human species. Some scientists have calculated that more than 90% of our actions are shaped by dopamine.

In the light of this understanding of dopamine's role in driving our behaviour, let's re-visit the Youth in Iceland programme discussed in Chapter 2. This programme did not promote the idea of staying safe by cutting back on alcohol, drugs, and smoking, i.e., asking young people to forgo their dopamine hits. Such messages are not effective, as has been demonstrated by dozens of campaigns that employ rational arguments. Instead, the programme was built around activities that offered young people an immediate and relevant dopamine hit by engaging with alternative activities. It was based on the understanding that young people will seek a dopamine hit, and that what we can do to shape their behaviour is to deliver this hit in a more socially desirable way.

Cortisol – Stress and Fear

When we are stressed our brain releases cortisol, the stress hormone. We may think of cortisol – also called the 'stress neurotransmitter' – as the opposite of dopamine. When the brain releases cortisol, we experience anxiety and stress. This was extremely useful in a natural hostile environment. When humans found themself in a dangerous situation, their brain released cortisol which set off the stress response – a surge of adrenaline to help them fight harder or run faster; growing fat cells in their abdomen to store more food if they killed an animal; and a reduction in their metabolism rate to free up more energy to deal with the challenge they were facing.

The stress response is truly amazing and boosts the chances of survival in a stressful situation where our behaviour may be a decisive factor in our survival. But when we experience chronic rather than acute stress, i.e., stress that continues over a long period of time – a much more likely scenario in the modern world than facing down a wild beast – our hardwired stress response is not only useless but damaging, as it can lead to major health issues. What saved human life in a natural, hostile environment makes life miserable for many people and kills some of us today. Chronic stress has repeatedly been identified as a cause of severe mental and physical illnesses that can lead to death (Mayo Clinic, n.d.).

Importantly, just as the brain releases dopamine when we *expect* a positive experience, it releases cortisol when we fear a negative one. Thucydides, evaluating the Peloponnesian War between Athens and Sparta over 2,000 years ago, concluded that the conflict resulted from growing Athenian power instilling fear in Sparta. For some 50 years before the war, Athens developed slowly but consistently into a major Mediterranean power. It was Sparta's fear of an emerging superpower that sparked the war. In 'The Thucydides Trap', Graham Allison examined whether the same dynamic would apply to the relationship between the United States and China. He identified 16 hostile episodes in history when an established power feared a challenger might eventually become a major competitor or adversary. Twelve of those cases ended in war (Graham Allison, 2018).

A nation or tribe wanting to extend their territory or capture some desirable assets (such as oil) may start a war. Even more often, however, we see wars ignited by fear. Consider some of the conflicts we have seen since the Second World War. The Vietnamese communists didn't represent a direct threat to the United States. However, the Americans were afraid of a growing communist world. No chemical weapons were found in Iraq, yet the Americans feared their existence (or, perhaps, they worried about their oil supply). The Taliban in Afghanistan did not constitute a serious threat, but the Russians and later the Americans feared they could become a safe haven for Muslim extremists. In the present war in Ukraine, Russia have claimed they invaded a peaceful country because of their fear that NATO and the EU (and, by association, the Americans and the neo-liberal West) were using Ukraine as a staging post for forms of incursion into Russia. Fear rather than actual events even drives the emerging cold war between China and the United States.

In summary, fear is a significant driver of human behaviour. In contrast to dopamine, which we seek, we want to avoid cortisol.

Think back to the Blue Mountains Hydroelectric Scheme example in the previous chapter. Frequent electricity blackouts occurred when Australia boosted its manufacturing capability during the Second World War. In other circumstances, nobody would have given electricity any thought; blackouts turned electricity supply into an immediate and personal threat. It is unlikely there would have been much public support for the scheme, and the major cultural disruption it represented, if people had not experienced these inconveniences. Their behaviour, in this case support for the Blue Mountain scheme, was driven by a desire to avoid cortisol.

Oxytocin

Oxytocin[3] is a hormone, but it can also act as a neurotransmitter involved in establishing social bonds. It is released by the brain during positive social interactions or the process of falling in love. It starts the reward circuit, which in turn releases dopamine, producing a sensation of well-being which also promotes pro-social behaviour and interpersonal affective bonds. Oxytocin reduces fear and anxiety and increases feelings of well-being and relaxation. It is common for people who feel under threat to move closer together, but it is equally common for this to happen when individuals feel they can collectively exploit major opportunities.

Feeling part of a group of people who share the same values or views about a particular topic or issue reassures us by providing a sense of being part of something bigger, of not being alone. If we want to engage people, we need to create initiatives and events that bring them together, creating a tribe that not only feels attached to our cause, but also allows for these followers to interact with each other, triggering repeatedly strong dopamine hits.

The Sports Club Recife is a prime example of how belonging can influence actions. We know that there was a lack of organ donations prior to the Club's engagement. The feeling of belonging to the Club was so strong and important that its members were prepared to engage with an unrelated initiative.

Belonging seems to be gaining in importance today as a growing number of people feel a sense of loneliness. In fact, loneliness is becoming a major issue in today's society. Cigna has tracked the incidence of loneliness in the United States for many years. Before the pandemic, 54% of adults felt lonely, but this had increased to 61% by 2020 (CIGNA, n.d.).

Mirror Neurons

There is something strange about us humans. When we are exposed to news that hundreds or even thousands of people are suffering or have been killed, we may feel sympathy and may even be outraged. Nevertheless, our emotional involvement is typically superficial and quickly recedes. But if we are confronted with a specific person who is suffering from injustice or a terrible fate, we are much more likely to feel strong emotions. The media, not-for-profit organizations, advertisers, and others use this fact by featuring the stories of individuals to generate stronger emotional involvement in broader issues. But why do we respond like this?

It happens because our brain contains mirror neurons which allow us to feel what others feel. This explains why we can feel anger, fear, or joy when we watch a character in a movie experiencing these emotions. Our mirror neurons don't react in the same way when we are faced with more abstract news about some misfortune or disaster to which we cannot relate, such as the millions of refugees languishing in detention camps. This is an important point: A rational approach may suggest that we should highlight how overwhelming a problem is by quoting the large number of people that are being affected. But a highly emotional story about an individual or small group is more likely to impact how people feel about the challenge and is thus more likely to lead to action.

Oxytocin, Serotonin, and Endorphins

Humor is one of the most effective short-term facilitators of escape. It triggers dopamine, oxytocin, serotonin, and endorphins. Humor brings down our defenses and makes us feel closer to the person that delivers it. Oxytocin, also referred to as the empathy hormone, reduces fear and anxiety behaviour and increases feelings of well-being and relaxation. It also can generate a feeling of relatedness. Serotonin controls our moods and makes us feel content and happy. Finally, endorphins trigger feelings of pleasure which reduce anxiety, create a feeling of safety and, by doing so, lift our mood. As a result of the release of these neurotransmitters, we feel less stressed as our heart rate, blood pressure and muscle tension – all signs of stress – are eased.

You may recall from Chapter 2 that humor was an important ingredient in wage negotiations in Austria. In fact, it also played a role in the negotiations with Russia to lift Austria's occupation, which led to a withdrawal of the occupation forces in 1955. One example of this is the bet between the Russian president and the Austrian representative as to who would be able to breed a bigger pig. It may seem trivial, but these humorous bonds can do much to soften conflicts.

Time to Revisit Our Success Stories One More Time...

Let's summarize: Humans are not naturally good, so telling them the right thing to do is unlikely to be effective. Meanwhile, a high degree of familiarity stunts our emotional response. We get used to change that takes place incrementally over a longer period. More specifically, our expectations are shaped by past experiences, and once we expect the next Climate Conference to deliver unsatisfactory results, a superpower to invade another country, or yet another political scandal, we don't feel any outrage that might have previously triggered us to act. Instead, we accept that this is just 'the way things are'. Ukraine may be an exception to the complacency of the Western world, due to Ukraine being "like us" rather than a country in the African continent or in South America, and also because Russia is threatening to use nuclear weapons which can reach other European, North American, and, indeed, the shores of all other countries.

What can we do to shape behaviour, given these discouraging facts?

While the case studies presented in the second chapter span more than seventy years, they all have something in common: Success was due to aligning solutions with how the brain works.

In the case of the Blue Mountains Hydroelectric Scheme, we do not doubt that Australia changed its immigration policy and accepted non-British European immigrants only because Sydneysiders experienced blackouts and shortages of fresh produce that required intensive agriculture. There was a feeling of clear and present danger – a challenge that affected people directly and immediately. More specifically, the Snowy Mountains project was sold as a means of reducing cortisol (We won't need to worry about electricity and food shortages any longer!), with some dopamine elements coming onstream as it evolved (We are embarking on the biggest engineering project the world has seen! We are heroes!).

Here is a related story: When a particularly ferocious bushfire killed many people in rural Australia, the government established an early warning system. They would send SMS and email messages to all inhabitants of small country towns when an extreme fire risk developed. Residents would then leave their homes and go to a dedicated secure area to wait out the fire. A survey of residents showed that 60% would leave when they received such a warning. The other 40% had various reasons for saying they would stay, mainly because they wanted to defend their homes against the bushfire. Eventually, the day came when the first warning had to be sent due to extremely high bushfire risk. And what happened?

Less than 1% left.

Market research conducted after this discouraging event showed that many residents went outside as soon as they received the warning. But they decided that there was no problem because they could not see or smell smoke. As far as they were concerned, there was no immediate, personal danger.

But the story does not end there. The researchers also found that several of the male residents kept looking up and down the street to see if anyone else was leaving. They were worried that they might end up being the only one acting in compliance with the safety instructions and that, should there be no fire, they would end up being the butt of jokes. It may be hard to believe that anyone would risk their family's life for fear of being ridiculed by their mates. Still, for many men their standing amongst their peers was a paramount consideration and, importantly, it was of immediate concern and had a high degree of relevance. At the same time, the bushfire without any sign of smoke was neither immediate nor did it appear relevant.

Would it be possible to design the bushfire safety programme to be more effective? One option would be to identify the informal community leaders and assign them roles, such as responsibility for clearing their street or managing the logistics at the safe spot. This would have given them a position of importance, delivering a dopamine hit and an incentive to make the programme work. Again, shaping behaviour would require aligning the approach with how the brain works.

A bit of history helps us understand the Austrian case example. From the end of the Second World War in 1945 to 1955 Austria and its capital city, Vienna, was divided into four districts, each occupied by one of the Allied forces – the Soviet Union, the United States, the United Kingdom, and France.

The occupation lasted so long because the Western allies apparently believed Russia intended to annex Austria; in order to prevent this threat, they also decided to stay. Whatever the reason, the end of the occupation was largely attributed to Nikita Sergeyevich Khrushchev, who served as the First Secretary of the Communist Party of the Soviet Union from 1953 to 1964. Khrushchev came from peasant stock and worked in a mine when the revolution that changed his life took place in Russia in 1917. He was one of the great statesmen of his time.

The Austrian Bundeskanzler, Julius Raab, and his predecessor, Leopold Figl (who had spent time in concentration camps and was destined to be killed when the Allied Forces freed Austria), both reportedly had good relations with Khrushchev. This personal relationship contributed much to Russia's decision to lift the occupation (Republic of Austria, 1955). Notably, when Figl formed the first government in 1955, he established a coalition with the Socialists and Communists, *despite his People's Party holding an absolute majority*.

This brief history is a great example of how memories shape behaviour. Austrians' shared history led to a strong belief in collaboration, in working together rather than fighting factions. This spirit is also evident in the wage negotiations our case study was based on and it persisted as long as the 'old guard', the people who shared the memories of the war and its aftermath, were in power. It follows that looking at history can sometimes be an essential step towards finding a solution to a challenge we face today.

From a neurotransmitter perspective, it's worth noting that the Austrian wage negotiations were designed to deliver dopamine (every party has something positive to announce) while simultaneously reducing the interference of cortisol (a friendly, non-combative atmosphere, sharing food, and feeling part of something important).

The Iceland example is quite straightforward: We are hardwired to seek dopamine hits. Smoking, eating, dangerous activities, drugs, and other unwanted behaviour generate a dopamine hit. We need to provide people with alternative, desirable ways of getting their dopamine, rather than telling them to give up the behaviour that currently provides it.

In Brazil, the Recife Soccer Club case example shows the power of the hardwired drive to belong. This is again a straightforward example that hardly needs more interpretation.

Finally, we have Germany's approach to phasing out coal mining. There are some cultural elements at work here, not unlike the Austrian case example. But it again illustrates the power of belonging. The fact that the country is taking care of the miners and others who are disadvantaged due to the mine closures shows a strong, shared responsibility to all Germans and a responsibility to look after those who – by no fault of their own – find themselves in difficult circumstances. This may also be one of the factors that led Germany to invite one million refugees to settle there, something no other country has ever done. And just as Germany managed the integration of East and West Germany into a single country following the fall of the Berlin Wall, it also managed to absorb these refugees without a significant economic setback despite many experts predicting Germany would go bankrupt. In summary, Germany's closure of coal mines combined a dopamine strategy (a vision of a better future) with a cortisol-reducing component (nobody will be left behind).

At the time of writing these words, there has been an interesting development in the United States: Senator Joe Manchin III, a Democrat from Western Virginia, has declared that he will not support President Biden's Build Back Better initiative (Jessica Bursztynsky, 2021). This will render this initiative dead in the water, as every Democrat needs to vote in favour to pass the legislation. The future of coal mining lies at the heart of the argument. The Senator has a personal, commercial interest in coal mining, but claims that he is simply trying to protect the coal mining industry. But something unexpected has happened: the United Mine Workers Union has publicly declared that they are supporting Biden's initiative, which calls for incentives to attract renewable energy companies to coal-producing areas and the funding of retraining schemes. An interesting battle is shaping up: a self-interested politician versus an enlightened labour union. In the context of our discussion, this just shows that it is possible to go through major transitions as long as there is a shared commitment to support those who will be disadvantaged by the change.

Unfortunately, we don't find support for a transition from coal to renewables in Australia where decisions are shaped by another hardwired driver of behaviour: The drive to compete. The coal-mining areas are in marginal electorates. The government was keen to show their support for the locals, and a new mine will bring jobs. Not many, but some. The latest Climate Conference, COP-26, has just finished in Scotland. Unsurprisingly, the Australian Prime

Minister has shown no willingness to scale down emissions; rather, he has kicked the can down the road. Money talks: Who cares about Australia's Pacific Nations' neighbours seeing their islands disappearing in the ocean? That's not Australia's concern.

These case examples show that human behaviour is not random, but some key drivers determine how we act. This is, of course, a complex matter.

Neurotransmitters such as dopamine and cortisol are key factors. These neurotransmitters shape what we seek and value, such as belonging to a group of people. As we will see in a later section, the mix of neurotransmitters is quite different for males and females and it changes during key live events such as giving birth or experiencing menopause, as well as due to ageing. These neurotransmitters activate some basic drivers such as a need to belong, explore, compete, or seek rewards. There is also evidence to support the idea that there are neurological bases for institutions such as religion which fulfill many of our basic needs for feeling secure.

We are not aware of what's going on in our old brain, which is why it is often referred to as our nonconscious mind. Yet it controls how we feel and drives much of our actions because it is more powerful than our new, conscious mind and it likes to take shortcuts. It has developed over many generations while humans lived in a hostile natural environment and is very much acting like we still find ourselves there.

But whatever is driving our actions, it stands to reason that we will be more successful in shaping human behaviour if we align our solutions with the way the human mind works.

Having gained a deeper insight into what is shaping our dominant response to major challenges and the drivers of behaviour shaping them, we are now ready to address specific challenges.

Notes

1 Let us quickly add that there are, as always, exceptions. Jacinta Ardern or Angela Merkel come to mind, and there are undoubtedly many more, though surely still in the minority.
2 We are excluding immediate, reactive behaviour, such as touching something hot and the brain releasing a neurotransmitter to warn you to remove your hand.
3 https://www.verywellhealth.com/hypothalamus-8646628.

References

Allison, Graham (2018) *Destined for War: Can America and China Escape Thucydides' Trap?*, UK, London: Scribe; quoted by Zanny Minton Beddoes, Manichean and Messy, in *The World Ahead 2022*, The Economist, pp. 13–14.

Booth, William, and Adam, Karla, Boris Johnson, Wife and Chancellor Amongst Those Fined for Downing Street Lockdown Parties, *Washington Post*, April 12, 2022. https://www.washingtonpost.com/world/2022/04/12/partygate-fines-downing-street/

Bregman, Rutger (2021) *Humankind. A Hopeful History*, Boston, Massachusetts, United States: Little Brown & Company.

Brunt, Martin Coronavirus: How Criminals are Exploiting the COVID-19 Pandemic to Scam the Public, *Sky News*, March 24, 2020. https://news.sky.com/story/coronavirus-how-criminals-are-exploiting-the-covid-19-pandemic-to-scam-the-public-11962897

Bursztynsky, Jessica. Sen. Joe Manchin Says He Won't Vote for Biden's Build Back Better Act, Potentially Killing the Social and Climate Bill, *CNBC*, December 19, 2021. https://www.cnbc.com/2021/12/19/sen-joe-manchin-says-he-wont-vote-for-bidens-build-back-better-act.html

CIGNA (n.d.) *Loneliness and Its Impact on the American Workplace*, Cigna, www.cigna-com/docs/about-us/newsroom/studies-and-reports/combatting-loneliness/loneliness-and-its-impact-on-the-american-workplace.pdf

Cutting Edge Training Developed the Human Brain 80,000 Years Ago, June 21, 2011, www.esciencenews.com

De Gabriele, Louis, and Abela, Jamine COVID-19 – And Criminal Exploitation, mondaq, March 30, 2020.

Pandemic Profiteering: How Criminals Exploit the COVID-19 Crisis, Europol, December 7, 2021. https://www.europol.europa.eu/publications-events/publications/pandemic-profiteering-how-criminals-exploit-covid-19-crisis

State Treaty and Neutrality, Republic of Austria (1955), www.parliamnt.gv.at

Stress Symptoms: Effects on Your Body and Behavior (n.d.), Mayo Clinic, www.mayoclinic.org

Trafton, Anne, *How Dopamine Drives Brain Activity*, McGovern Institute, MIT, April 1, 2020 and *How Does Dopamine Drive Our Behavior?*, www.crewbase.net

4

ADDRESSING GLOBAL CHALLENGES

We have deliberately chosen four quite different challenges to explore here – the refugee crisis, homelessness, food insecurity, and technologies eliminating jobs. The refugee problem requires a strategy at the country level; homelessness is primarily a city-level problem we can find even in wealthy countries; food insecurity largely – albeit by no means exclusively – afflicts the developing world; while the impact of technologies on work will initially affect primarily the developed world, but then evolve into a major force of change around the world. Can we find an approach that will close the Commitment Gap on diverse challenges such as these?

We are not suggesting that the solutions we discuss here are the best ones, or even that they are feasible. We don't have the firepower of a community of like-minded people, or the resources governments or corporations can bring to bear. We need collective brainpower to develop sound strategies, and collective action to have the necessary impact. We won't win all our battles, but hopefully enough of them to avoid the dystopian future we often see in books and movies.

Challenge 1 The Refugee Crisis

> It always seems impossible until it's done.
>
> Nelson Mandela

We have seen how Australia's Snowy Mountains Hydroelectricity Scheme led to that country abandoning its British-centric immigration policy to accept refugees from various other countries. This was a significant step forward. Unfortunately, Australia has since taken two steps backwards.

DOI: 10.4324/9781003477167-4

Today, Australia is spending big to detect and capture refugees. These unfortunate people are attempting to reach its shores to escape war, racial and religious persecution, hunger, and disease. Australia has deliberately chosen to label these refugees 'illegal immigrants' (a misnomer, as people applying for refugee status are not immigrants and applying for refugee status is not unlawful) and places them in offshore prison-like facilities where they often linger for many years, sometimes until the end of their lives.

The cost of these prisons is astronomical. According to figures compiled by the Refugee Council of Australia, the annual cost of overseas detention *per person* was $573,000 in 2018.[1] How much would it cost to accept these refugees, offer them English classes, a cultural introduction to Australia, healthcare, and job training? It would obviously be a whole lot less. But the government's decision is not based on economics; it is purely political.

The Australian Human Rights Commission's report on the use of force in immigration detention reports excessively harsh, punitive, and degrading conditions. This includes the unnecessary handcuffing of women, children, people in wheelchairs or with mental illness and no history of violent behaviour, and other people needing medical care. Handcuffing has become a routine practice for transfers between centres and also, alarmingly, for off-site medical appointments. Handcuffs are even applied when refugees are transferred for torture and trauma counselling and at consultations where the very purpose of the appointment is to investigate issues such as wrist and arm injuries.

The UN Human Rights Council and other organizations concerned with human rights issues have repeatedly criticized Australia's overseas detention centres and the indefinite detention of some refugees and asylum seekers (Ben Doherty, 2018). In 2018, the refugee with the longest stay in a detention centre had been there for nine years (Refugee Council of Australia, 2022)! How could a country purporting to be a respectable citizen of the world end up treating refugees in the most inhuman ways? The answer: The politics of fear, used by politicians of both major political parties – Liberal and Labour.

Given that the politicians' approach is based squarely on what they believe is palatable to the electorate, the challenge is to change from the bottom up – that is, beginning with the electorate. As long as citizens buy the politicians' unfounded stories of refugees causing risks to their country and their own way of life, we will see no change in policy.

Germany has arguably done more than any other country in accepting refugees. At the time of writing, the fact remains that the European Union is paying Turkey €3 billion to keep millions of refugees in detention camps (Leo Cendrowicz, 2015). And let's not forget the role the United States, Russia, and other countries have played in killing hundreds of thousands and making millions homeless, hungry, and cut off from social, educational, or medical support, without any intention of rebuilding the countries they have destroyed.

The Big Picture

The United Nations High Commissioner for Refugees (UNHCR) estimated that, by the middle of 2023, the number of people forcibly displaced stood at more than 110 million, with over 36.4 million refugees (Refugee Council of Australia, 2023). This is the highest number in recorded history. To put this in perspective: The total population of Canada and the United Kingdom combined is approximately 107 million. Imagine all the Canadians and British people in refugee camps. This is the size of the challenge, and it is growing, due to conflicts such as the Russian invasion of Ukraine and Israel entering the Gaza Strip, where at the time of writing the Israel Defence Force have killed more than 40,000 civilians, leading to a massive number of refugees seeking safety elsewhere.

Many refugees are held in camps where there is a lack of running water, with proper sanitary facilities, regular food (let alone healthy food), medical care, education, and other basic needs either limited or not available at all. Many have been traumatized by their experiences and are in urgent need of support services. Importantly, approximately half of all refugees – i.e., 40 million people – are children growing up in a desperate environment, with little opportunity to participate in schooling. They lack the diverse socializing opportunities that would benefit their development, and have very little chance of a fulfilling and meaningful future ahead of them.

There are occasional glimpses of greatness: For example, Germany made an enormous contribution when it took in one million refugees in 2015, but even this tremendous effort made only a small dent in the global numbers. Of course, if Germany's leadership had inspired other countries, it would have had much more impact, but that did not happen. Instead, some world leaders predicted that as a result of this action Germany would experience massive difficulties and that their Bundeskanzler, Angela Merkel – one of the most successful leaders on the world stage – would lose the next election. It didn't happen that way, but there is still no appetite elsewhere to follow the German example (Sekou Keita and Dempster, 2020).

Even worse, however, the Christian Democratic Union (CDU), the party Merkel led for several decades and which is today a major opposition party, is now trying to win votes by promising to send refugees to third countries for the processing of asylum applications, either African countries such as Ghana or Rwanda, or non-EU European countries such as Moldova or Georgia. It is hard to believe that this is the party that had accepted one million refugees under Merkel and which calls itself the *Christian* Democratic Union (*Christlich-Demokratische Union*)!

The UK announced the same 'third country' policy in 2022, but in 2023 the Supreme Court ruled that the government policy was unlawful (British Red Cross, n.d.). We must be thankful that the judiciary is keeping some politicians in check.

Unfortunately, this is how political expediency works: rather than attempting to educate the public, it is much easier – and presumably more successful – to simply reinforce any trends by announcing policies that fuel them.

If we can't deal with the refugee problem today, can we at least ensure that the numbers don't grow any further? This leads to the question of what causes the displacement of people in the first place. But, of course, we all know the answer: wars, persecution, and civil unrest are top of the list, followed by economic hardship. According to Brown University's Costs of War Project, by the year 2000 the United States' 'War on Terror' accounts for the displacement of 37 million people (Watson Institute, n.d.). It should have been more aptly named the War *of* Terror.

In summary, there are no easy solutions to the refugee challenge, which is no surprise: if there were easy ways of dealing with it, we wouldn't have a problem today. But perhaps we need to look at some of the reasons why tackling the refugee challenge head-on is so unpalatable, and what could be done to address these issues.

Killing Myths

There is a myth suggesting that terrorists are hidden amongst refugees.

It is reasonable to expect that some refugees will become radicalized as they forced to stay in substandard camps or are homeless for extended periods. There is, of course, a simple way of managing this risk: Provide constructive assistance and resettle refugees within a reasonable time. If there are well-founded expectations that a new life can begin in, say, a year, it would be much easier for refugees to bear the burden of their miserable life in the interim. Uncertainty is a massive problem: Refugees typically have no idea when they will be allowed to resettle somewhere or if they will die in their refugee camp. They are totally without power, without information, without reasonable expectations. In that sense they are worse off than prisoners who know how long their incarceration will last.

However, the idea that terrorists hide amongst refugees, often advanced by politicians and the media, is ludicrous. A trained terrorist is valuable and will not come in a rickety boat to some country to spend years in a detention camp, having no idea when a country might accept them, nor which country they would end up in. Terrorist organizations are not short of funds. Terrorists are more likely to arrive by air as tourists or business travellers. Unfortunately, myths like these have the power to turn citizens against refugees.

Another myth is that refugees are 'jumping the queue.' This is a view often held by immigrants who have gone through the tedious bureaucratic processes to finally get accepted by a host country. They are right, of course, that refugees did not go through these bureaucratic channels. The truth, however, is that they could not. Many of them are from war zones or countries devastated by war. Think about Syria, with villages bombed, homes and facilities destroyed. There is a lack of medical care, food, and water. Your chances of applying for a visa to immigrate to a country are nil. In fact, reaching somewhere where you could apply is impossible; the chances of survival while waiting (if you could apply) are even less.

The idea that these refugees are attempting to jump the queue is cruel and born of ignorance and arrogance.

What is true, however, is that terrorism is being bred in refugee camps where refugees are treated badly, isolated from the world, denied medical and social care, and given no idea when they might be allowed to move on towards a new life. Imagine spending many years in a refugee camp like this. Is there a risk that you develop a hatred for the people who mistreat you so badly? Would you become an easy recruit for a radical movement?

Belonging is a key behavioural driver. Belonging reduces cortisol, the 'stress neurotransmitter', and – when members of the group support each other – delivers dopamine. Furthermore, there is conditioning: as mentioned earlier, some 40% of refugees are children. Their view of the world, their values, ambitions, attitudes, and opinions are shaped to a large degree by this early life experience. Given all that, how likely is it that refugee camps are a breeding ground for radical movements? What do you think?

But here is the irony: keeping refugees, including many families with children, in these camps for years under horrific conditions is itself the potential driver of radicalism. The answer to this challenge is not to refuse refugees entry into a country or to deport them to some far-away place, but to start managing the refugee problem rather than just placing them in camps where horrific conditions lead to either resignation or hate.

There is a parallel to today's vaccination challenge: Wealthy countries are storing Covid-19 vaccines, and Pfizer allegedly made the supply of its vaccines dependent on countries signing agreements preventing them from shipping excess quantities of vaccines to the developing world. For Pfizer, this would be a sound, though inhumane commercial approach. In any case, the developed countries will ultimately pay for this as the virus mutates in communities with low vaccination rates, and the mutated strains then travel around the world. Not a problem for the pharmaceutical industry as they benefit from the need for additional booster shots or new vaccine variations, but a problem for the many thousands that may die or suffer debilitating after-effects due to the developed world just ignoring developing countries.

This situation is well known and has been reported in both the popular press and academic and research literature (Karim, 2022). For example, the Royal Society of Arts (RSA) published research findings that demonstrated that countries such as the UK and US had put significant resources into developing vaccines which enabled them, some may say fairly, to be the first countries to receive the vaccines as these became available. However, the ability of rich countries to first develop and then take prime place in the rollout of vaccines has left poorer countries, especially those from the African continent, at the back of the queue and largely unvaccinated.

The scale of the problem of the unequal distribution of vaccines in Africa is demonstrated in a December 2021 report by the African regional office of the World Health Organization (WHO). They noted that of the roughly 10 billion doses of Covid-19 vaccine that had been given out by the time of their report, less than 10% of people from lower-income countries had received a dose and just 3% of Africans had been given a shot: Immunization rates were nearly eight times higher in rich countries when compared with countries form the African continent. At the time of the WHO report, around 11 billion doses of vaccine had been produced which could have immunized about 40% of the global population. The report authors predicted that at the current rate of production, 70% of the global population could be given a shot.

Vaccines for Covid are designed to reduce the consequences of the pandemic on life, health, and also on health services in individual nations. The most vulnerable need to be inoculated first, such as the elderly and those with health issues, but the distribution of vaccines to people based on their need is not possible when rich countries build up disproportionately large stocks of the vaccines.

However, there is also a recognition that, as Karim claims, "no one is safe until everyone is safe" (Karim, 2022). Viruses, by their very nature, mutate and in developing countries, and especially in the African continent, unvaccinated populations are breeding new Covid variants that then spread to the developed world. National boundaries do not deter the spreading of viruses; whilst countries only address the needs of their own citizens (a short-term and narrowly focussed interest) the pandemic will continue. This situation can only be addressed by richer countries sharing more doses of vaccine with poorer countries. Moreover, in all countries, the scourge of fake news and misinformation must be addressed and countered. One initiative to facilitate the more equitable distribution, called COVAX, was developed out of a partnership of organizations with the WHO. The aim was to give 2 billion vaccine doses to 190 countries by the end of 2021 with 92 low-income countries receiving doses for free. However, COVAX achieved the delivery of only 1 billion doses due to rich countries striking their own deals with the vaccine manufacturers.

The same is likely to happen with refugees: Ignoring them if they don't represent an immediate and personal danger will eventually lead to the radicalization of a few inmates in these camps, and eventually we may well see terrorist attacks. The sad end of this story is that politicians will present this as proof that refugees are dangerous when, in fact, it was the lengthy detentions and inhumane conditions that caused the radicalization. The same principle applies here as it does to Covid-19 vaccinations: no one is safe until everyone is safe. Terrorists can strike anywhere at any time; the investment in detecting and preventing attacks is much more expensive than the cost of treating refugees humanly and helping them to rebuild their lives.

What Can Be Done?

The Australian Snowy Mountains case study demonstrated how a practical problem – in this case, electricity and food shortages – was addressed by accepting displaced people into the country as a much-needed labour force. Are there major concerns today that could be addressed by taking in refugees?

US President Biden's trillion-dollar Bipartisan Infrastructure Law/Infrastructure Investment and Jobs Act has passed the House in November 2021 (Jim Probasco, 2022). Of course, the intention is to create a massive number of jobs for Americans as funds are released to state and local governments to fix bridges, the railway network, and roads, and to expand internet access. The Act addresses infrastructure needs created by fifty years of neglect. For example, 173,000 miles of the nation's highways and major roads and 45,000 bridges are in poor condition. There are thousands of miles of sewage and water pipes that need to be replaced, and levees and dams that need to be repaired or reinforced. Anyone who would like greater insights into the scale of neglect should visit the website of the American Society of Civil Engineers or the US Army Corps of Engineers.

Infrastructure development will initially have a negative impact on climate change, but will reduce emissions in the long term due to electrification and efficiency gains, improved road and rail infrastructure, and more efficient cities. The funds have been set aside for major infrastructure works, and this list of fund allocations will give some idea of the enormous scale and scope of this infrastructure programme:

- Roads, bridges, and major infrastructure projects: $110 billion
- Bridge repair, replacement, and rehabilitation: $40 billion
- Major projects too large or complex for traditional funding programmes: $16 billion
- Transportation safety: $11 billion
- Public transport modernization: $39 billion
- Passenger and freight rail: $66 billion
- Broadband upgrade: $65 billion
- Port infrastructure: $17 billion
- Airports: $25 billion
- Nationwide network of electric vehicle chargers: $7.5 billion
- Rebuilding the electrical grid: $65 billion
- Water infrastructure: $55 billion

- System resilience, i.e., protecting the system from droughts, floods, and cyberattacks: $50 billion
- Rehabilitating land: $21 billion.

This is undoubtedly an ambitious infrastructure programme, probably only challenged by China's investments in infrastructure. But China has a much larger reservoir of workers than the United States. In the US, the speed of progress will be hampered not only by bureaucratic processes and a lack of experience in managing massive infrastructure projects at state and local government levels, but also by a lack of labour.

We have focused on the United States because the Infrastructure Law has provided us with a comprehensive overview into the country's needs. But the demand for major projects of this nature is by no means limited to the United States. And it is not only neglected infrastructure maintenance that requires addressing; climate change requires massive *new* infrastructure development. Sea levels have already started to rise; as this continues, a significant number of people will be affected. Climate change is turning infrastructure development into a major challenge.

Rising sea levels arguably allow us to predict the impact of climate change with somewhat greater accuracy than other climate change-related events such as storms and wildfires. A study published in Nature Communications suggests that by the year 2100 many people will find that the land they live on is below sea level (Scott A. Kulp and Strauss, 2019). More specifically:

- 10 to 50 million each in China, India, and some southeast Asian countries
- 1 to 9 million each in the United States, Brazil, Germany, the UK, Egypt, and other countries
- 500,000 to 999,000 each in Canada, Mexico, Colombia, Peru, the USSR, Finland, Australia, and several others.

Our thinking might run into a problem at this stage: we know that people are unlikely to react to something likely to happen only in the long term. The threat is not immediate, and therefore not relevant. Consequently, we can expect property prices to drop in areas that will be affected; other than that, however, we are unlikely to see a significant impact for two or three decades.

However, while the risks of climate change may not be seen as an immediate threat, we can expect the infrastructure programme to lead to a multitude of commercial opportunities and jobs – and these are of an immediate nature. Importantly, it will take decades to construct sea walls, dams, and other necessary fortifications. Just look at Venice (admittedly not a prime example for efficiency). Venice started to build sea walls and floodgates in 2003 and only some 18 years later had their first trial run. The London Flood Gates took an entire decade to complete, and they are only dealing with a single river. If we want to protect the coastal areas of major cities like Miami, New Orleans, New York, Hong Kong, or Shanghai, we need to start work on massive infrastructure projects soon.

The important point is that the American administration has recognized this. The Infrastructure Bill makes all the difference. We can hope for some serious action in the United States, and this will spark similar moves by the European Union and elsewhere around the world, driven by an element of competition, another hardwired driver of behaviour. Governments will fund programmes, but their execution will require a vast workforce. A bonded immigration scheme that allows refugees to immigrate in exchange for working on a public benefit programme for a couple of years could make a real difference to both refugees and the advancement of major infrastructure projects.

Importantly, many of the required services could be delivered in a highly cost-efficient and effective way because of the large number of refugees that would live and work together. Centralized health and welfare services, language and cultural education classes, and organized activities with locals to help assimilation and integration could be undertaken in a highly cost-efficient way. Importantly, refugees would share the same challenges and opportunities with each other, creating bonds that will last a lifetime. They would get paid and would be able to save some money to have a better start once they have completed their time on the project they signed up for.

Allow yourself to think big: Imagine an international programme allowing countries to post their requirements and the conditions they offer, allowing refugees to apply for specific programmes that suit their skill sets and preferred countries. Refugees with valuable skills or experience would naturally have an advantage. Still, there is no doubt that there would be many jobs where a brief training period is sufficient to develop the required competencies.

Such a programme would deliver a win–win for participating countries, our ability to manage the impact of climate change more effectively, and for refugees who would get a chance to start a new life. But, of course, nothing is ever without risks.

How is this solution aligned with the way our brain works? Just as in the Blue Mountains example, we are changing from a goal that seems to have little relevance to one that is highly desirable. Most people clearly don't care about refugees or see them as endangering their culture and way of life. It would take generations to disabuse people of these misperceptions. In fact, it may never happen as it is unlikely that rational arguments and fact impact on emotionally held beliefs.

However, climate change has now reached a stage where it is causing significant damage. For an ever- increasing number of people, it is becoming an immediate and relevant danger. According to a Washington Post article, for example, more than 40% of Americans live in counties hit by climate disasters in 2021 (Sarah Kaplan and Tran, 2021).

Our concept addresses their goal to deal with this challenge, i.e., to protect them from the worst climate change will bring. This reduces stress (cortisol) and, once there are some success stories, we can expect dopamine to fuel the desire for more action on this front. Refugees are not any longer seen as taking jobs and destroying a way of life, but rather helping to preserve the quality of life by managing the impact of climate change, the world's greatest challenge.

Managing Risk Factors

There are potential pitfalls we need to manage:

The potential for exploitation

Arguably one of the most significant risks is the abuse of refugees. While they would be a low-cost way of addressing a significant challenge, it is vital that they are treated fairly and not exploited. Otherwise, the whole programme will fail. Here are some risk factors:

- abuse, keeping refugees in substandard accommodation, without proper healthcare, and withholding payments
- lack of support services such as health and welfare services, education for children, language, and cultural programmes for adults
- lack of support in settling refugees once they have completed their indentured period.

There is a risk in assigning refugees to private sector operators with a single-minded profit objective. At the same time, it is unlikely that the public sector can effectively manage the massive infrastructure development projects required. This suggests that we need to establish effective and comprehensive independent, ongoing supervision and auditing. Ideally, this organization's board should bring together refugee advocacy representatives, infrastructure experts (with no direct financial interests), health and welfare specialists, and union and government representatives.

Competing with locals for jobs

The programme will never get off the ground if locals feel that the refugees are taking their jobs. Of course, it is designed to avoid this, i.e., to address the massive infrastructure challenges that will be beyond the existing labour pool. However, we need to ensure that locals don't fear that refugees might reduce their own employment opportunities. An obvious solution would be to regularly publish available jobs and allow only domestic workers to apply in a first round (e.g., with a two-month lead time). Any jobs left at the end of this round would be offered to refugees.

Politicization

At the time of writing, the United States Administration is currently considering the establishment of a Civilian Climate Corps, a federally funded initiative aimed at employing tens of thousands of young people to fight climate change by cutting trails, building roads, and solidifying America's infrastructure. Republicans are opposing it simply because a Democratic Administration advanced it.

In countries where partisan politics rules, we must be careful to introduce the programme via independent organizations and only after a public survey shows broad support. Naturally, explaining the programme, the safeguards taken to protect local jobs, and the benefits to the locals in managing the impact of climate change or simply improving outdated infrastructure will need to be part of such a survey.

In conclusion to this discussion, we return back to our earlier Declarative Mapping Sentence (DMS) (Figure 1.1) to analyse our commentary and offer potential ways forward (see Chapter 1, for a more detailed analysis about the DMS). Figure 4.1 outlines a DMS representation for the refugee challenge.

FIGURE 4.1   THE HACKETT MODEL: DECLARATIVE MAPPING SENTENCE REPRESENTATION FOR THE REFUGEE CHALLENGE (CHALLENGE 1)

A typical response in the face of a refugee challenge is dependent upon a person's:

Stage 1

greater awareness
neutrality towards
lesser awareness

that the refugee challenge exists which then leads to belief that the challenge is of:

Stage 2

critical importance to sustainable development
somewhat important
of little importance

requiring the person to believe that their actions may be:

Stage 3

effective i
ineffective

in terms of the refugee crisis, and that there is a/an:

> *Stage 4(a)*
>
> possibility of success
> impossibility of success
>
> *for individual refugees, and the*:
>
> *Stage 4(b)*
>
> possibility of success for countries/regions

Challenge 2 Homelessness

> When life gets hard, try to remember: the life you complain about is only a
> dream to some people.
>
> Anonymous

Homelessness statistics are difficult to gather. Many people are homeless for only a limited time. In some countries they can register, but not everyone does. However, even if the figures we have overestimate the real problem, they are daunting: The World Economic Forum reported that 150 million people were homeless worldwide (Homeless World Cup Foundation, n.d.; Downie et al., 2018). In the US, more than 500,000 people were homeless on a single night, and Euronews reported that at least 895,000 people were homeless in Europe in 2020 (US Department of Housing and Urban Development, n.d.; Isabel Marques da Silva, 2023). Women are arguably more at risk of violence than men, and US statistics show that nearly 30% of homeless people are women. As terrible as homelessness is for anyone, it is shocking that many children are in this situation.

There are many reasons for homelessness, including domestic or family violence, unemployment, alcohol or drug use and abuse, financial difficulties, eviction or housing affordability stress, divorce or relationship breakdown, mental health issues, physical health issues, ending incarceration or army service with nowhere to go, and relationship or family breakdown (Humanrightscareers, n.d.). In some places the Covid-19 pandemic had a positive impact on homelessness – for example, temporary orders restricting landlords' ability to evict tenants who can't pay the rent, or schemes to house the homeless during lockdowns – but it has also made homelessness riskier in other ways due to lack of access to vaccinations or medical help.

The impact of being homeless is often devastating. By the end of 2021, 227,000 households across the United Kingdom were experiencing homelessness (Crisis, 2021). The average age of death for people experiencing homelessness is 45 for men and 43 for women. People sleeping on the street are almost 17 times more likely to have been victims of violence. More than one in three people sleeping rough has been deliberately hit or kicked or experienced some other form of violence while homeless. Given these har-rowing experiences, it may be unsurprising that homeless people are over nine times more likely than the general population to take their own life.

We would expect most people to agree that having homeless people on city streets is undesirable. Of course, some will feel like this because it makes them uncomfortable to be confronted with homelessness, or they fear a negative impact on property prices or tourist spending. Others believe that we have a responsibil-ity to address this challenge if we, as a community, can afford to do so. But what is a person to do? It is easier to rationalize that there will always be homeless people and we just must accept it, or that it is too expensive to deal with the problem. In other words, ignore the problem or deny that it can be addressed.

While the refugee crisis is global, and therefore needs cooperation between different countries for it to be addressed, the homelessness challenge is a local-ized one that is – unsurprisingly – prevalent in cities. The US Department of Housing and Urban Development reported that, in 2020, the three most signif-icant problem areas were New York City with just under 78,000 homeless peo-ple, followed by 64,000 in Los Angeles City & County, and 11,700 in Seattle/ King County (US Department of Housing and Urban Development, n.d.).

A Local Problem

Here is the real issue: Some cities have successfully addressed the homeless problem and have found that the cost of doing so tends to be lower than the cost of homelessness. This is the tragedy of homelessness: It is unnecessary to have all these unfortunate people living in the streets. We can afford to address this problem, but few people know or care. If they did, then maybe they would lobby their city administration to take decisive action.

The Housing First approach has been the stand-out solution when it comes to addressing homelessness. It demands offering permanent housing to the home-less as quickly as possible, followed by essential services as soon afterwards as practical. This puts Housing First in sharp contrast to the widely used temporary 'band-aid' approach, i.e., providing sheltered or transitional housing.

Interestingly, Housing First was developed in the 1990s and has success-fully addressed homelessness in several cities, yet these represent only a small minority of all the places that could benefit from adopting this policy. In any case, the idea is to remove the main obstacle that stands in the way of getting the homeless started on a path to a better life: to make having a home a fun-damental human right. There are no waiting times, nor do homeless people need to fulfill certain criteria to qualify.

Case management services support the previously homeless who now find themselves in a permanent home. This allows professionals to help these res-idents to deal, when required, with past trauma, a lack of self-confidence, substance abuse, and other issues. In the United States, a pioneer in Housing First, tenants pay 30% of their income towards rent.

While this sounds quite simple, in practice, it is not. It is a complex undertaking that requires funding. However, there is room for optimism as cities that have embraced this solution have reported positive outcomes concerning the number of homeless as well as their city budgets. It appears that the Housing First programme – which may seem unaffordable considering the cost of housing – saves money by reducing the cost of other services such as welfare, police, the justice system, and health services. Here are some reports:

- Research in Seattle, Washington, showed that providing housing and support services for homeless alcoholics costs taxpayers $4 million less in its first year compared to leaving the homeless on the streets and spending on police and emergency services (Larimer et al., 2009).
- The Utah Division of Housing and Community Development reported a 72% decrease in overall costs since enacting the plan in 2005 (Utah Housing and Community Development, 2014).
- A research study at the University of Northern Carolina reported that a housing project for the chronically homeless called Moore Place saved the county $2.4 million (SHNNY, n.d.).
- The US Department of Housing and Urban Development reported that the number of chronically homeless individuals living on the streets or in shelters dropped by an unprecedented 30%, from 175,914 people in 2005 to 123,833 in 2007, with the Housing First approach seen as a contributor to this reduction. (And yes, the Housing First concept has been around that long.)
- Another report showed that the Housing First approach resulted in a 66% decline in days hospitalized, a 38% decline in emergency room visits, a 41% decline in EMS (Emergency Medical Services) events, a 79% decline in days in jail, and a 30% decline in police interactions (Fortune, 2013).
- A report compiled by two public health academics provides a comprehensive assessment of the cost-effectiveness of Housing First (Wright and Peasgood, 2018). The report highlights potential savings, but also points out that any long-term estimates are sensitive to the cost of supported accommodation and the cost of providing Housing First support.

The Housing First approach has also found its way into European countries, with the Czech Republic, Finland, France, and the UK adopting programmes of this nature in some cities.

The CAUF Society (caufsociety.com), dedicated to reducing suffering due to homelessness, claims that four cities have just about eliminated homelessness by adopting a Housing First scheme. The city that was the first to get there is Vienna, where housing projects for homeless people have been developed in downtown locations. The Austrian Federal Government is spending more than $700 million on construction and habitation for subsidized housing. While this is a very small share of the Federal Budget, it has helped limit homelessness in Austria.

Helsinki, Finland, is another successful city. Helsinki has adopted the Housing First principle of unconditional housing for all. Many cities require homeless people to meet specific criteria and conditions before becoming eligible for permanent housing. By contrast, Helsinki offers housing unconditionally, based on the belief that having somewhere to live must be addressed first before people can focus on the problems that may have gotten them into that situation.

Of course, it won't be possible to solve the homelessness problem totally and absolutely. Some homeless people will refuse support because of mental health, drug abuse, or other reasons, and a small number may choose to live like this. However, we would make this world a much better one if we enabled most homeless people to rebuild their lives again. Yet homelessness does not seem to be a high priority in most places.

In some countries, the barriers may be financial in nature: the savings are for services primarily provided by a Federal Government, while the costs are carried by City Councils. Of course, this problem could be overcome if both parties were committed to addressing homelessness.

Conclusions

There is no doubt in our minds that there is a Commitment Gap when it comes to homelessness. We know that the problem can be solved. There is evidence showing that the solution is affordable.

The local community typically elects city office bearers. They are often not career politicians, nor experts in city management. Rather, they tend to represent a voter group with particular interests or demands, such as developers, property owners, businesspeople, or people who want their cities to be greener. I expect many of these elected representatives do not know that there are proven solutions to homelessness and that this is likely to lead to a net financial gain for their city.

The city administrators are the key advisers. They are more likely to be aware of Housing First and other solutions to homelessness. But they may see these options as disruptive because they would require the abandonment of the traditional service model serving the homeless and its replacement with a new approach. We are not suggesting that the administrators ignore this option, but they may focus too much on the administrative and management implications rather than the improved quality of life for the homeless and improved city ambiance.

The result is denial: the key advisers and decision-makers don't like to face up to the problem. They would rather ignore it or, when asked, pretend it does not exist or that it cannot be solved. Meanwhile the citizens electing them are unlikely to know that there are well-tested, affordable solutions.

An obvious line of attack is to promote the case examples that demonstrate the positive impact of Housing First in cities across the world such as Vienna, Helsinki, and Salt Lake City. This effort should include conferences and presentations to key decision-makers, but also social media activities that would raise awareness that there are workable solutions amongst the all-important voters who will decide on the next leadership team when elections come up.

This first wave of activity targets ignorance. This will be enough to get the city over the line in some instances. More action, however, may be needed in other places. We will present a range of strategies in a later section.

In conclusion to this discussion, we return back to our earlier DMS (Figure 1.1) to analyse our commentary and offer potential ways forward. Figure 4.2 lays out a DMS representation for the challenge of homelessness.

FIGURE 4.2 THE HACKETT MODEL: DECLARATIVE MAPPING SENTENCE REPRESENTATION FOR THE HOMELESSNESS CHALLENGE (CHALLENGE 2)

A typical response in the face of *a homelessness challenge* is dependent upon a person's level of:

Stage 1

greater awareness of hunger and food-insecurity challenge
neutrality towards the hunger and food-insecurity challenge
lesser awareness of hunger and food-insecurity challenge

that the homelessness challenge exists which then leads to belief that the challenge is of:

Stage 2

critical importance to sustainable development
somewhat important
of little importance

where the person to believes that their actions may be:

Stage 3

effective in terms of ...
ineffective
in terms of the homelessness challenge, and that there is a/an:

Stage 4(a)

possibility of success
impossibility of success

for individual persons impacted by the homelessness challenge, and the:

Stage 4(b)

possibility of success for countries/regions.

Let's close with a quote that sums up today's situation:

These things become the norm: that some homeless people die of cold on the streets is not news. In contrast, a ten-point drop in the stock markets of some cities is a tragedy.

– Pope Francis

Challenge 3 Hunger & Food Insecurity

Naturally, hunger statistics are not consistent as it is difficult to make an accurate assessment in areas where hunger is prevalent. But the overall picture is consistent: hunger is a major global problem:

The World Health Organization reports that around 735 million people faced hunger in mid-2023, compared with 613 million in 2019. Considering these developments, it is clear that the UN Development Goal of ending hunger by 2030 will not be reached (World Health Organization, 2023).

According to the UN, some 820 million people suffered from hunger in 2019, up from 811 million the year before. The UN report notes that income equality is rising in many countries where hunger is also on the rise, suggesting that a small segment of the population is capturing much of the resources while others are left behind. When we broaden our focus and look at food insecurity, we find that 17.2% of the world's population, i.e., 1.3 billion people, lacked regular access to nutrition and sufficient food. Finally, concerning children, the UN report notes that no progress has been made in reducing low birth weight since 2012. At the same time, we find that overweight and obesity continue to increase in all regions, including among school-aged children (UN Department of Economic and Social Affairs, n.d.).

The 'Action against Hunger' movement states (Action Against Hunger, n.d.):

- 3.1 billion people cannot afford a healthy diet
- 783 million people suffer from hunger around the world
- 1 in 10 people suffer from hunger
- 80 million more women than men were hungry in 2022
- 85% of people facing hunger crises live in conflict-affected countries
- 13.6 million children suffer from severe acute malnutrition
- 2 million children die every year from malnutrition
- 45% of all child death worldwide are from hunger and related causes.

Again, we would like to put this in perspective: the European Union has a population of approximately 450 million and the United States one of 340 million. Imagine hunger affecting all these people!

To make matters even worse it has become clear that the war in Ukraine will also have a negative impact on food supplies, as Ukraine and Russia are major suppliers of grain, sunflower seeds, and other agricultural products (ING Report, 2022).

While these factors are hopefully of a medium- rather than a long-term nature, there are several factors contributing to food shortages in the long term, including climate change, land use, and existing agricultural practices. We will focus on two major drivers of these shortages: food waste and agricultural subsidies.

Food Waste

> 'There are nearly a billion undernourished people in the world – but all of them could be fed with just a fraction of the food that rich countries currently throw away.'
>
> Tristram Stuart, food waste activist

We don't think of food waste when we occasionally throw out some food at home. Unfortunately, it is a huge problem. The UNEP Food Waste Index shows that, globally, households waste some 569,000,000 tons of food, food service companies like fast-food chains 244,000,000 tons, and retailers a further 118,000,000 tons. This adds up to 931,000,000 tons of wasted food a year (UNEP, 2021).

Given that there are an estimated 820,000,000 people experiencing hunger over long periods, sometimes their whole lives, this amounts to more than a ton of food for every one of them. That's just over 3 kg per person per day. Hunger problem solved. Only, of course, it isn't solved as we can't get the wasted food to the people that need it.

There is also associated hidden waste, specifically the water used to produce food that ends up being discarded. The Water Footprint Network reports that more than 15,000 liters of water are required to produce one kilogram of bovine meat, just over 9,000 liters for one kilogram of nuts, 8,700 for sheep/goat meat, just under 6,000 liters for one kilogram of pig meat, 4,300 for chicken meat, 3,200 for eggs, 1,600 for cereal, 1,000 for milk, just under 1,000 for fruit, and 300 for vegetables (Statista, n.d.). These are global averages. Of course, in some regions it doesn't matter because water is plentiful; others, however, water is – or will be – in scarce supply due to population growth and the climate crisis.

Remember the CIA report we mentioned in the history of the climate crisis: It raised the potential for war due to food and water shortages. Just as nations compete for oil and other resources today, in future they will compete for food and water when they experience shortages. At this stage, conflicts are limited mainly to arguments about river water use. For example, Ethiopia, Sudan, and Egypt are competing for the water of the Nile; Turkey, Syria, and Iraq are in conflict over the Euphrates-Tigris basin; Afghanistan and Iran argue about the Helmand River and the Hari Rud; and let's not forget the dam conflicts in the Mekong River basin (EU Science Hub, 2018).

Killing Myths

The average household not only wastes a significant amount of food but is often unaware of how much a bit of waste here and there adds up to. Statista reports the following about consumers in different developed countries (Patrick Wagner, 2018):

- In Great Britain, consumers believe they waste 5% of food, while the actual percentage is 15%
- In Canada, the respective percentages are 10% and 21%
- In the US, 15% and 24%
- In Switzerland 5% and 18%, and so forth.

Clearly, a lack of awareness is the first major hurdle when addressing the scale of food waste.

While the waste per household may be small, the problem turns into a considerable challenge when considering the number of households. The UNEP Food Waste Index Report 2021 estimates that China's households are wasting 64 kilograms per capita, or a total of nearly 92 million tonnes (UNEP, 2021). The figures for India are 50 kg per capita for a total of just under 69 million tonnes, the United States comes third with 59 kg per capita and 19.4 million tonnes, followed in order by Japan, Germany, France, the UK, Russia, Spain, and Australia based on total food waste. The waste per capita is by far the greatest in Australia, at 102 kilograms per capita, followed by France at 85 kilograms and then the UK and Spain at 77 kilograms.

The big question is: Could eliminating this waste provide more food for the hungry? After all, households can't just send leftovers destined for the dustbin to some faraway country to feed the hungry. However, there are two outcomes we could pursue:

First, an impact could be made on local people suffering from food insecurity. What if you bought a tin of some food every time you have saved by not wasting food and made this extra tin available to a foodbank or a charity that looks after the homeless or other disadvantaged people?

Second, part of the solution may be an increased awareness of the ramifications of food waste, particularly the resources wasted in growing, processing, and distributing food that ends up being discarded. In recent years, in fact, there have been campaigns in many countries aimed at reducing domestic food waste. These campaigns have tended to focus on the money wasted by the average household on discarded food – not an unreasonable strategy, given that it makes the problem more personally relevant.

But to address the overall food waste issue, we need to go further back up the supply chain.

Getting Supermarkets and Processors to Waste Less Food

Why have so many companies invested into becoming – and proving that they have become – sustainable organizations or offering sustainable products? The simple reason is that the public's stated interest in sustainability encouraged them to invest in meeting the perceived demand for sustainable enterprises.

Once sustainability was on the agenda, it triggered a secondary driver: CEOs and other senior managers or owners want recognition for leading the sustainability battle. Sometimes this has its roots in a company wanting to beat a competitor to a sustainability accreditation; at other times, it is a more personal competition between CEOs or a way for a CEO to capture the limelight; finally, in some instances, these efforts may be driven by the values held by key decision-makers.

Whatever the specific drivers may be, the important point is that consumers – collectively – have a lot of influence on what companies do. Suppose a large number of consumers showed a deep interest in food wastage and rewarded companies that had taken steps to minimize this malpractice. In that case, we could expect companies to follow the path they took with respect to sustainability.

Significant public interest in food waste would lead the food processing and service industry and retailers to minimize their waste. As with sustainability, there could be an accreditation scheme that would provide a reliable assessment of the company's performance. And an offset scheme – again as with climate change – would allow them to balance their unavoidable waste with an investment into food production or supply in regions where hunger is rampant.

In our exploration of food waste, we should not forget the many positive actions that have already been taken by some of the parties causing the problem. Here are some examples:

- Dutch supermarket chain Albert Heijn's AH Overblijvers programme is expected to save around 4.5 million kg of food waste in the Netherlands annually. The scheme allows people to buy boxes of products near or past their best-before dates directly from Albert Heijn for drastically reduced prices. Customers can reserve boxes via an app, choosing from themes such as 'vegetarian', 'bakery', and 'surprise'. Albert Heijn's platform shows that these purchasing behaviours are moving into the mainstream and will soon be an expected option for many (Alex Strang, 2022).
- Aldi is offering affordable eco-friendly options to customers, such as distributing surplus food and products, launching a dedicated eco-store, and raising awareness of food waste amid surging grocery costs. In another scheme, Lidl and Too Good to Go offer people vacuum-sealing equipment for their food to help maintain freshness and prevent unnecessary waste at home (John Firth, 2022).

- Across the Netherlands, Wasteless is implementing its AI-informed pricing system to reduce grocery store food waste. This tech allows for effortless participation in a sustainable practice for shoppers, as well as helping to keep prices low. The system works by analyzing the stock data and age of products and then automatically displaying markdowns on shop shelves – ensuring the product is sold at an ideal price for both buyer and seller. Wasteless claims to reduce 50% of food waste and increase business revenue by 20% (Raini Kenyon, 2023).
- The Too Good to Go food app connects consumers with retailers who have food that is nearing its expiry date, providing bags of randomly selected items for a significantly reduced price. This is a great example of allowing consumers to take positive action (reduce food waste) while also gaining a tangible, immediate benefit (inexpensive food). The app has teamed up with a total of 36,000 stores worldwide and boasts a total of 1.84 million app downloads across Europe (Roisin Lanigan, 2020).

While food waste remains a challenge, there is a hidden driver of food insecurity that also needs addressing: agricultural subsidies.

Agriculture Subsidies

'When we give a subsidy, the benefits to the public ought to exceed the benefits to the company. When it doesn't, that's our definition of corporate welfare.'

John Kasich. US politician

Many people don't know that agricultural subsidies contribute significantly to global food shortages (Damian Carrington, 2019). The nations accounting for the greatest amounts spent on subsidies are the United States, the EU, and Japan in absolute terms, while the Philippines, Indonesia, and China allocate the largest share of GDP in the form of subsidies. Importantly, in developed countries, these agricultural subsidies typically don't support the consumer, i.e., make food more affordable; rather, they boost the producers' profits.

Agricultural support is defined as the annual monetary value of gross transfers to agriculture from consumers and taxpayers arising from government policies that support agriculture, regardless of their objectives and economic impacts. In 2019, US net farm income reached $111.2. billion. This includes the US government's Farm Subsidy Payments of $22.6 billion to farmers and owners of farmland. In 2000, US government payments accounted for 45.8% of total net farm income (USA Facts, 2023). Meanwhile, the EU has approved income support for farmers of €188 billion for the period 2023–27 (European Commission, n.d.).

Undoubtedly, some of these subsidies support hardship cases or help farmers when natural disasters trigger a loss of income. But much of it simply boosts the attractiveness of this sector to investors as government subsidies boost the return on investment. Moreover, in the US and other developed countries, farms are no longer run by families of local people; rather, they are in the hands of large corporations, meaning that subsidies are supporting industry which may or may not support local communities.

This means that rather than investing in, say, agricultural development in Africa, where massive productivity increases could be achieved and millions of people could escape a life of food insecurity, investors stay with the mature – but heavily subsidized – agricultural sectors of Europe and the US. Importantly, these subsidies also depress prices, making it very difficult for farmers in developing countries to compete in international markets.

The problem is, of course, a political one: How do you ween this sector off these subsidies without losing votes? Many of these farmers and corporations have significant political power. Unsurprisingly, they are more interested in their businesses than in addressing the subsistence conditions subsidies create in some developing countries or the persistent challenge of a billion people going to bed hungry.

Reducing farm subsidies is not an attractive proposition for many citizens. However, remember the closure of coal mines and coal-powered electricity stations in Germany? The wealthy United States or the EU could phase out subsidies that do not support struggling farmers while allowing others to adjust over time with programmes that would allow farmers to phase out of the industry or lift farm productivity. Of course, this would not be an easy road to take. But we do need to make some significant changes if we want to deal with global food insecurity. The important point is that we need to accommodate and support those who are caught by these necessary adjustments.

In conclusion to this discussion, we return back to our earlier Declarative Mapping Sentence (Figure 1.1) to analyse our commentary and offer potential ways forward. Figure 4.3 lays out a DMS representation for the challenge of hunger and food insecurity.

FIGURE 4.3   THE HACKETT MODEL: DECLARATIVE MAPPING SENTENCE REPRESENTATION FOR THE HUNGER AND FOOD INSECURITY CHALLENGE (CHALLENGE 3)

A typical response in the face of *a hunger and food-insecurity challenge* is dependent upon a person's level of:

Stage 1

greater awareness of hunger and food-insecurity challenge
neutrality towards the hunger and food-insecurity challenge
lesser awareness of hunger and food-insecurity challenge

that the hunger and food-insecurity challenge exists which then leads to belief that the challenge is of:

Stage 2

critical importance to sustainable development
somewhat important
of little importance

where the person to believes that their actions may be:

Stage 3

effective in terms of ...
ineffective

in terms of the hunger and food insecurity challenge, and that there is a/an:

Stage 4(a)

possibility of success
impossibility of success

for individual persons impacted by the hunger and food insecurity challenge, and the:

Stage 4(b)

possibility of success for countries/regions

Challenge 4 Technologies Eliminating Jobs

> The development of full artificial intelligence could spell the end of the human race... It would take off on its own, and re-design itself at an ever-increasing rate. Humans, who are limited by slow biological evolution, couldn't compete, and would be superseded.
>
> Stephen Hawking, BBC

Of course, it all depends on who you are listening to. Organizations that offer solutions that replace workers or assist with their implementation, their major shareholders and lobbyists, or politicians who want to attract high-tech firms tell us that new technologies are here to help us, to eliminate boring and demeaning jobs so we can focus on whatever is more important, and generally to lift the quality of work we do. Importantly, they don't eliminate the number of jobs as they create new ones.

Others suggest that anyone who is not using the latest technologies is likely to suffer the consequences. As consultant and Harvard Business School Professor Karem Lahkami said: *'AI won't replace people, but people who use AI will replace people who don't'* (Karem Makhani, 2023).

We need to address two questions to develop a position on the likely impact of technologies on jobs: First, will technologies destroy more jobs than they create? Second, if they do, what are the solutions open to us to avoid massive unemployment?

Will Technologies Destroy More Jobs Than They Create?

Here is a small sample of recent reports on job losses:

- A research study suggested that ChatGPT could eliminate 51% of marketing jobs, while lifting productivity by 74% (Anne-Gaëlle Sy, 2023).
- According to a report by Goldman Sachs, up to 300 million full-time jobs could be impacted in the next decade. The report suggests that approximately two-thirds of jobs in the US and Europe are exposed to some degree of AI automation (Ben Wodecki, 2023).
- A survey by Sortist Data Hub found that 26% of European software and tech companies are planning to cut jobs as a direct result of ChatGPT (Ben Wodecki, 2023).

Unsurprisingly, given the stake some organizations and individuals have in advancing the use of technologies to replace labour, there are also plenty of reports suggesting that new technologies will create new jobs (which is true) and some even claim they will generate more jobs than they destroy (which is not true).

It is a fact that technologies are replacing jobs. However, this does not mean that technological progress is a bad thing. Some jobs simply should not be undertaken by people because they are too hazardous.

Other jobs are critical and need to be done even when people can't do them. Think about the pandemic and the importance of continuing with healthcare, basic supplies, vaccinations, and more. When service providers may get infected and can't deliver, it is important to have a technology solution that still can.

There are also widely used traditional technologies that are associated with a high mortality rate. Motor vehicles fall into this category. The Center for Disease Control and Prevention (CDC) reports that, globally, 1.35 million people are killed every year in road accidents, i.e., nearly 3,700 a day (Centers for Disease Control and Prevention, 2023). This is about the same as the population of San Diego dying every year on the road! In addition, some 50 million people are injured – more than the population of Spain! Autonomous vehicles could save millions of lives and avoid tens of millions of disability cases. However, on the downside, they would also destroy the livelihood of an untold number of professional drivers.

In other words, we are not looking at a black & white picture. Let's look at some of the myths that have been put forward to assure us all that the technological revolution is either not a challenge or that, if it is, there are simple solutions. These myths prevent decision-makers from taking decisive action while keeping citizens at large content and happy. But they are not based on the reality of what is already happening, or is likely to happen, as the technological revolution unfolds:

Myth 1
This is just another industrial revolution – it will work itself out like all the others!

It is not just another industrial revolution. For the first time, we have developed technologies that can outperform us with respect to specific cognitive abilities and develop new technologies and solutions without human input or intervention. Moreover, there is an extensive infrastructure in place to facilitate the rapid deployment of many technologies.

Myth 2
It may happen – but not to me!

New technologies can match or improve human performance in previously unimaginable ways, offering both massive cost savings and, often, improved performance. Unless you belong to an elite group of highly skilled engineers or scientists, you will find that the technological revolution will eventually catch up with you.

Myth 3
It's just like another recession – and we know how to get through those!

We can reasonably expect that times will improve during an economic recession and, eventually, we will be 'back to normal'. Even a deep recession does not last forever. The technological revolution, however, will not stop. It will continue to change the way we do things and eliminate jobs that will never return, nor will they all be replaced by new jobs. The Luddites were right after all!

Myth 4
Employment figures are the best for many years – there is no problem!

Employment statistics (pre-pandemic) were healthy because many new jobs had been created in the design, development, and deployment of new technologies. In many applications, we see traditional solutions still being deployed in parallel to new ones that are developed, piloted, tested, and improved. For example, the legacy car manufacturers build electric vehicle plants while still keeping the production lines for internal combustion engine (ICE) cars going. However, once the traditional solutions are replaced, we will see massive unemployment.

Myth 5
Retraining will solve all problems!

Retraining will be important, especially in the earlier phases of the technological revolution, and every effort needs to be made to develop and offer access to retaining programmes. However, retraining will eventually diminish in impact as there will simply be few jobs to retrain for.

Myth 6
My government will look after me! The Universal Basic Income (UBI) will ensure everybody is okay!

The UBI is a much more dignified and efficient way of supporting unemployed people than current welfare schemes. However, to be affordable, it will have to be based on a payment well below the average wage; when many people end up with such a diminished income, we will see economies collapse.

Myth 7
We will all share the benefits of productivity gains – and live happily ever after!

There is simply no basis for believing that those who reap the benefits of the technological revolution will willingly share these with everyone else. There is ample evidence that many corporations today are unethical when it comes to paying their fair share of taxes, depriving governments of the tax revenues needed to provide services and support to those citizens who need it most. Just look at the tax avoidance by leading corporations and wealthy individuals today, and you can see the future!

Every one of these myths suggests either that there is nothing to worry about or that there is an easy solution for dealing with any negative impact the technological revolution might bring. While there is ample evidence that these myths are not grounded in reality, many people are happy to accept them as this eliminates the need to think – or even worry – about the challenge that lies ahead. And as long as citizens live in denial there is no pressure on politicians or industry leaders to develop strategies to deal with the developing job crisis.

The last two myth – the Universal Basic Income and the idea that we will all share the productivity benefits – deserve more consideration.

The Universal Basic Income Will Solve All Problems!

We want to shed some light on the Universal Basic Income, which is often presented as a solution to all sorts of problems, including technologies replacing jobs, homelessness, food insecurity, income inequality, and more. It has attracted much attention from the media, politicians, academics, authors, and the public. Essentially, the UBI promises everybody a regular income provided by the government without having to first qualify for such support. Several countries have conducted pilot programmes – or are in the process of doing so. And citizens, especially young people, are often only too willing to embrace the UBI. Of course, getting an additional monthly payment without having to do anything at all is an enticing prospect.

But is the UBI really the solution it appears to be?

The Good

Imagine you have lost your job. While you are looking for employment, you register for unemployment benefits, and this is where an undignified process starts.

Not only have you suffered the blow of losing your job, but you now also need to fill out endless forms, answer questions implying that you are just a shirker who doesn't want to work, you have to apply for a certain number of jobs every month to keep qualifying even if you are not a credible candidate, and if you are lucky enough to find some temporary work you may find your unemployment benefits cut, forcing you to re-apply once you have completed your temporary work.

Each country has a somewhat different unemployment scheme, but they tend to share one common feature: they are utterly demeaning. This is particularly bad as the vast majority of the unemployed find themselves in this situation through no fault of their own, suffer from their dismissal, and doubt their self-worth. Of course, we don't want people exploiting government services as this leaves less money for those who truly need them. But at what stage has the focus shifted from helping the many who deserve support to weeding out the few who abuse the system?

The UBI is, without doubt, a much more dignified way of supporting the unemployed. They don't have to fill out endless forms, attend interviews where they are treated like criminals or parasites feeding on the community, nor do they have to reject temporary work for fear of losing their benefits. They receive a sum of money with no questions asked. This provides them with a degree of security and an opportunity to chase employment opportunities without having to fret about a bureaucratic burden. Furthermore, the UBI is much less administration heavy. It would save a significant budget currently used to pay public servants administering unemployment benefit schemes that could be put to better use in helping those in need directly.

Let's have a look at a specific example: On 1 January 2017, Finland commenced with a pilot programme that provided 2,000 randomly selected unemployed Finns, aged 25 to 58, with a guaranteed monthly payment of €560 (US$587). This replaced the payment of their social benefits, and they received this income regardless of whether or not they found work.

To understand the reaction of people who were selected to receive the UBI, you first need to understand how Finnish unemployment benefits work: at the time of the scheme Finland had an unemployment rate of 8.7% (December 2016), with the unemployed receiving €697 per month in benefits, with small increases if they had dependent children. The unemployment benefits are taxable. Depending on age, the period for which these payments can be received is limited. Once a person ceases to be eligible, they can apply for labour market subsidies. The underlying idea is that recipients will make greater efforts to find work, even if it is just temporary. However, unemployment benefits are cut when a person finds work and reinstated only after a waiting period once that job has been finished. This means that unemployed people are not likely to show much interest in temporary work as they may be financially worse off compared to a situation in which they are not working at all. So much for unemployment benefits.

Now let's have a look at the Universal Basic Income: One key benefit is that the recipients of the Universal Basic Income do not need to prove that they are looking for work, they do not need to fill out forms or to attend any interviews. And they can undertake temporary paid jobs to supplement their income without the need to report on this and see their regular payments cut back. In other words, the Universal Basic Income delivers important benefits to unemployed people.

We know little about the impact of the UBI as the identity of the recipients is rightly being protected. Anecdotal evidence suggests that participants much prefer the UBI to unemployment benefits. There is no doubt that something like a UBI should replace unemployment schemes that degrade people and make them feel inferior. This is the Good – but, as the saying goes, nothing is perfect...

The Bad

The UBI – as a 'universal' basic income – has limitations.

First, the idea that everybody should get it is problematic. It means that most people who receive a monthly payment courtesy of the government don't need or deserve this additional income. Anyone living below or close to the poverty line should be supported and would be a worthy recipient. But higher-income earners should not benefit from a UBI as this would reduce the funds available to those in need.

The argument often put forward is that the UBI will, of course, be taxed. In other words, somebody in a high tax bracket will pay the highest tax rate on their UBI. Let's do a quick calculation to understand the impact of taxation: let's assume that there are 3 million people in the top tax bracket, paying a 33% tax on their UBI. In other words, they retain two-thirds of their UBI payments. With a monthly UBI of $2,000, we are looking at $48 billion that will be retained by recipients who don't need this cash injection! Of course, these figures will vary depending on the UBI payment and the top marginal tax rate, but the point remains: taxing the UBI is not addressing the problem of wasting money that is supposed to benefit those who need it.

Another problem is that even with those who deserve a UBI payment, the needs are not always the same. However, once we start to differentiate, we are back to filling out forms, attending interviews, and arguing our case. The productivity gains disappear, and dignity erodes.

These are the Bad aspects of the UBI. We believe the UBI's benefits out-weigh the disadvantages of a traditional unemployment scheme. However, given the impact a 'Universal' Basic Income would have, we favour limiting payments to unemployed people, replacing the often-undignified and administration-heavy unemployment schemes rather than providing an additional income to those who don't need it.

The Ugly

The beneficial aspects a UBI would deliver start to unravel when we look at a future with unprecedented high levels of unemployment. Unfortunately, this is a likely scenario, given that technologies are projected to replace jobs on a massive scale. But, unlike a recession when jobs are lost for a limited period until recovery sets in, the technological revolution will continue to facilitate the replacement of more and more jobs as it progresses. In other words, the question is not how many jobs will be replaced by technologies, but rather how quickly this will happen.

This is the background scenario. The UBI has been advanced by numerous parties as a means of managing the impact of the technological revolution. The argument is that we don't need to fear the technological revolution as there will be a safety net in place. But will the UBI live up to this role? To answer this question, let's return to the Finland pilot referred to earlier.

The average monthly income in Finland is €3,392 (€3,500 per month in the private sector) compared to the UBI of €560. The UBI pilot is with long-term unemployed people for whom it offers advantages over unemployment benefits. But what would happen if an ever-increasing number of displaced workers were to receive the UBI?

What would the economic impact be? For example, what would happen to consumer spending if 20%, 30%, or an even higher percentage of people end up living on a UBI that represents somewhere between 15 and 25% of average monthly earnings? They would have to cut back significantly on their purchases, sending businesses bankrupt, leading to further job losses – and an even larger percentage of the population relying on the UBI.

Unfortunately, increasing the UBI to a level approaching average income is not a viable option. Governments will be collecting less revenue from the business, corporate, and consumption-based taxes due to the increasing number of unemployed people and the resulting slowing of the economy. At the same time, we can expect a much higher utilization of welfare, health, and other government services. Government spending will go up while tax revenues decline. It will be impossible for governments to afford to offer a UBI pegged around the average income to unemployed people and certainly not to everybody.

In the succinct words of the report for the Institute for Policy Research by Dr. Luke Martinelli of Bath University: 'An affordable basic income would be inadequate, and an adequate basic income would be unaffordable.' (World Health Organization, 2023).

Governments may well put the UBI forward as a quick fix, given that the idea of receiving a monthly payment is attractive to most people. But imagine a large percentage of the population living on the low income delivered by the UBI with little prospect of ever gaining long-term, meaningful employment. They won't be able to plan for a better future, keep their aspirations alive, or save up for exciting holidays or essential major purchases. They are stuck with a subsistence income, resulting in a poor quality of life.

And then consider the impact on the economy and what it will mean to everybody – those who already have lost their job and others who are still drawing an income. The result will be a massive economic downturn as consumer spending collapses, companies cut back on costs, and more employees find themselves made redundant, contributing further to the growing mass of citizens who have incomes that may allow them to survive but not much more.

A Promising Strategy: Sharing the Available Work[2]

Debunking these myths frees us up to search for a solution that would be meaningful and effective. For example, we can reframe our discussion by focusing on the number of work hours available rather than the number of jobs. When we think about the impact of the technological revolution on jobs, we can only see jobs disappearing. But when work hours disappear, this does not mean that jobs must disappear too. What if we found a way to share the available hours? What if everybody worked less as technologies take over tasks, resulting in everyone still having a job?

Various pilot programmes have shown that people who work fewer hours are more focused, more productive, and enjoy better health. Some firms have even found that all the work gets done with the same number of people working fewer hours. However, shorter working hours have not always delivered the desired benefits. For example, the Swedish city of Gothenburg staged a two-year experiment at Svartedalen old people's home, cutting nurses' working hours to six hours a day while maintaining pay levels. The results showed that the nurses reported feeling healthier, with reduced sick leave and improved patient care. However, the City of Gothenburg decided that was unable to extend this programme as a result of its high costs. To cover the reduced hours for the 68 nurses it had to hire 17 extra staff at the cost of about $1.3 million. It was concluded that a six-hour workday would be unaffordable (Amanda Billner, 2017).

But what if the pilot had been undertaken in a manufacturing plant where robots would have taken over the additional work? An Austrian steel plant with an output of 500,000 tons would typically have employed some 600 to

800 workers. However, a new robotized plant employs just 14 people. What if the pilot programme had been conducted not in an aged care facility but in a steel plant, an assembly and production facility like Foxconn, or a warehousing operation such as Amazon's? It would have shown that it is possible to cut back work hours without creating a productivity or cost problem, as robots can complete much of the work done previously by humans at a lower cost.

What we are seeing is that in some organizations (like nursing homes) it is unaffordable to allow employees to work fewer hours at the same salary. In contrast, in others (say, plants that can utilize robots or office operations that can employ Artificial Intelligence), it is possible to do so and even to lift output while lowering unit costs.

At this point, let's introduce the late economist William Jack Baumol (Jack Baumol, 2012). He suggests that we need to look at a country's Gross Domestic Product (GDP) differently from what we do today.[3] His concept holds promise when searching for a way to turn the technological revolution into a positive development.

More specifically, Baumol suggests that a country's economy comprises of two sectors. The *Progressive Sector* is made up of goods and services that can benefit significantly from the productivity gains delivered by technologies. Meanwhile, goods and services that have only limited scope for productivity gains from the introduction of new technologies belong in the *Stagnant Sector*. Clearly, the Austrian steelworks we mentioned earlier is part of the Progressive Sector – with technological advances lifting a worker's productivity by a considerable multiple – while the nursing home in Sweden is part of the Stagnant Sector.

Now let's imagine a hypothetical country that generates half its GDP with industries in the Progressive Sector and half with industries in the Stagnant Sector. Over time, the Progressive Sector will deliver productivity gains, i.e., it will cost less to produce the same goods and services. The Stagnant Sector won't be able to gain these productivity benefits, and the cost of goods and services produced in that sector will not decline. This means that, over time, the Stagnant Sector will account for a growing share of the GDP. But – and this is the critical point – the standard of living won't decline: What it means is that *people will now pay less for goods and services the Progressive Sector delivers and more for goods and services the Stagnant Sector provides.*

We have already experienced this in many countries: we now pay much less than we used to as a proportion of income for cars, television sets, computers, kitchen, household equipment, packaged consumer goods, foodstuffs, and more – all goods that have benefited from productivity gains delivered by technologies. But we pay more for healthcare, personal services, and some kinds of education that do not benefit significantly from such productivity gains.

Let's assume for a moment that a country introduces a shorter working week across the board. This would force the Progressive Sector to utilize new technologies to stay competitive. When technologies are more productive than labour, we will see a decrease in costs despite reducing work hours. There will also be significant job losses as technologies take over more of the work.

The Stagnant Sector would fare differently: The shorter working hours would create a significant number of new jobs (in the Swedish nursing home example, the workforce grew from 63 employees to 80 employees, an increase in jobs of 27%). The cost of these services would go up, but consumers would be able to afford them because they would pay less for goods and services provided by the Progressive Sector.

In summary, everybody would be better off: Employees would work fewer hours, everyone would have a job, and the standard of living would not decrease. Of course, there would be a massive transition challenge as workers in the Progressive Sector would need to be retrained to find suitable jobs in the Stagnant Sector. This is no easy task: a steel worker may not find it easy to become a nurse! The point is that transitions are never without challenges, but we can avoid a much more damaging impact by pursuing them.

In conclusion to this discussion, we return back to our earlier DMS (Figure 1.1) to analyse our commentary and offer potential ways forward. Figure 4.4 lays out a DMS representation for the challenge of technologies eliminating jobs.

FIGURE 4.4   THE HACKETT MODEL: DECLARATIVE MAPPING SENTENCE REPRESENTATION FOR THE TECHNOLOGIES ELIMINATING JOBS CHALLENGE (CHALLENGE 4)

A typical response in the face of a technologies eliminating jobs challenge is dependent upon a person's:

Stage 1

greater awareness of *technologies eliminating jobs challenge*
neutrality towards *technologies eliminating jobs challenge*
lesser awareness of *technologies eliminating jobs challenge*

that the technologies eliminating jobs challenge exists which then leads to belief that the challenge is of:

Stage 2

critical importance to sustainable development
somewhat important
of little importance

where the person believes that their actions may be:

Stage 3

effective

ineffective

in terms of technologies eliminating jobs challenge, and that there is a/an:

Stage 4(a)

possibility of success

impossibility of success

for individual persons impacted by the technologies eliminating jobs challenge, and the:

Stage 4(b)

possibility of success for countries/regions

Notes

1 Updated data on the cost of holding refugees in Australian detention centers can be found at the University of New South Wales's Kaldor Centre: www.kaldorcentre.unsw.edu.au.
2 Adapted from Peter Steidl, *Time to Give a F*ck! The Technological Revolution and You*, 2019.
3 GDP here is defined as the value of all goods and services produced in a given country in a given period, typically a year.

References

10 Root Causes of Homelessness (n.d.) www.humanrightscareers.com

227,000 Households across Britain Are Experiencing the Worst Forms of Homelessness, Crisis, December 23, 2021. https://www.crisis.org.uk/about-us/crisis-media-centre/227-000-households-across-britain-are-experiencing-the-worst-forms-of-homelessness/

Agriculture and Rural Development. Income Support Explained, European Commission (n.d.), https://agriculture.ec.europa.eu/common-agricultural-policy/income-support/income-support-explained_en?prefLang=de

Baumol, Jack (2012). *The Cost Disease. Why Computers get Cheaper and Health Care Doesn't*. New Haven, Connecticut, United States: Yale University Press.

Billner, Amanda Swedish Six-Hour Workday Runs Into Trouble: It's Too Costly, *Bloomberg.com*, January 4, 2017.

Carrington, Damian $1m a Minute: The Farming Subsidies Destroying the World – Report, *The Guardian*, September 16, 2019. https://www.theguardian.com/environment/2019/sep/16/1m-a-minute-the-farming-subsidies-destroying-the-world

Cendrowicz, Leo Refugee Crisis: EU Pays €3bn to Turkey in Exchange for Help on Dealing with European Immigration, *Independent*, November 29, 2015. https://www.independent.co.uk/news/world/europe/refugee-crisis-eu-pays-eu3bn-to-turkey-in-exchange-for-help-on-dealing-with-european-migration-a6753861.html

Centers for Disease Control and Prevention, National Center for Injury Prevention and Control, January 10, 2023. https://www.federalregister.gov/index/2023/centers-for-disease-control-and-prevention

Isabel Marques da Silva At least 895,000 people are homeless in Europe as unfit housing conditions persist, new report says, *Euronews*, September 5, 2023. https://www.euronews.com/my-europe/2023/09/05/at-least-895000-people-are-homeless-in-europe-as-unfit-housing-conditions-persist-new-repo

Data provided by Statista (n.d.), www.statista.com

Doherty, Ben UN Body Condemns Australia for Illegal Detention of Asylum Seekers and Refugees, *The Guardian*, July 8, 2018. https://www.theguardian.com/world/2018/jul/08/un-body-condemns-australia-for-illegal-detention-of-asylum-seekers-and-refugees

Downie, M., Gousy, H., Basran, J., Jacob, R., Rowe, S., Hancock, C., Albanese, F., Pritchard, R., Nightingale, K., and Davies, T. (2018) *Everybody In: How to End Homelessness in Great Britain*. London: Crisis, www.crisis.org.uk

Federal Farm Subsidies: What the Data Says, USA Facts, October 5, 2023. https://usafacts.org/articles/federal-farm-subsidies-what-data-says/

Firth, John Aldi Raises Food-waste Awareness Amid Surging Costs, Canvas8, October 19 2022. https://usafacts.org/articles/federal-farm-subsidies-what-data-says

For Details on Potential Water Disputes see: *Global Hotspots for Potential Water Disputes*, EU Science Hub, October 16, 2018. https://phys.org/news/2018-10-global-hotspots-potential-disputes.html

Fortune, Sue, Pathways to Housing First Model Adapted for Use in the Canadian Context, October 2013, Saskatchewan. https://web.archive.org/web/20140221175538/http://www.saskatoonhealthregion.ca/your_health/documents/HousingFirstModel-SueFortin.pdf

Global Homelessness Statistics, Homeless World Cup Foundation (n.d.) www.homelessworldcup.org

Information on the Policy to Send Asylum Seekers to Rwanda, British Red Cross (n.d.) https://www.redcross.org.uk/get-help/get-help-as-a-refugee/information-on-the-policy-to-send-asylum-seekers-to-rwanda#:~:text=On%2015%20November%202023%2C%20the,seeking%20asylum%20in%20the%20UK

Kaplan, Sarah, and Tran, Andrew Ba More Than 40 Percent of Americans Live in Counties Hit by Climate Disasters in 2021, *Washington Post*, January 5, 2021. https://www.washingtonpost.com/climate-environment/2022/01/05/climate-disasters-2021-fires/

Karim, A. A. (2022) All For One: Nearly Two Years into the Pandemic We Have Failed to Address Vaccine Inequalities, *Royal Society of the Arts Journal* (1), 20–23. https://www.thersa.org/rsa-journal/2022/issue-1/feature/all-for-one

Keita, Sekou, and Dempster, Helen Five Years Later, One Million Refugees Are Thriving in Germany, Center for Global Development, December 4, 2020. https://www.cgdev.org/blog/five-years-later-one-million-refugees-are-thriving-germany

Kenyon, Riani Wasteless Uses Dynamic AI-pricing to Reduce Food Waste, Canvas8, January 18, 2023.

Kulp, Scott A., and Strauss, Benjamin H. (October 2019) New Elevation on Data Triple Estimates of Global Vulnerability to Sea-level Rise and Coastal Flowing, *Nature Communications*, 10(1). https://doi.org/10.1038/s41467-019-12808-z

Lanigan, Roisin *Too Good To Go: Saving Restaurant Food from the Bin*, Case Study, Canvas8, February 25, 2020. https://www.canvas8.com/library/case-studies/2020/02/25/too-good-to-go

Larimer, Mary E., Malone, D. K., Garner, M. D., Atkins, D. C., Burlingham, B., Lonczak, H. S., Tanzer, K., Ginzler, J., Clifasefi, S. L., Hobson, W. G., Marlatt, G. A. (2009) Health Care and Public Service Use and Costs Before and After Provision of Housing for Chronically Homeless Persons with Severe Alcohol Problems, *Journal of the American Medical Association*, 301(13), 1349–1357.

Latest Research Shows around 735 Million People Currently Facing Hunger, Compared to 613 Million in 2019, World Health Organization, July 12, 2023. https://worldstatustoday.com/blog/latest-research-shows-around-735-million-people-currently-facing-hunger-compared-613-million

Makhani, Karem AI Won't Replace Humans – But Humans With AI Will Replace Humans Without AI, *Harvard Business Review*, August 4, 2023. https://hbr.org/2023/08/ai-wont-replace-humans-but-humans-with-ai-will-replace-humans-without-ai

Moore Place Permanent Supportive Housing Evaluation Study, Research, SHNNY (n.d.) shnny.org

Over 820 Million People Suffering from Hunger; New UN Report Reveals Stubborn Realities of 'Immense' Global Challenge, UN Department of Economic and Social Affairs (n.d.) https://www.un.org/es/desa/over-820-million-people-suffering-hunger-new-un-report-reveals-stubborn-realities

Probasco, Jim Infrastructure Investment and Jobs Act, *Investopedia*, January 14, 2022, www.investopedia.com

Refugee Council of Australia *How Many Refugees Are There in the World?* October 29, 2023. https://www.refugeecouncil.org.au/?s=April+5%2C+2022+How+many

Statistics on People in Detention in Australia, Refugee Council of Australia, April 5, 2022, www.refugeecountil.org.au

Statistics provided by the US Department of Housing and Urban Development (n.d.). https://data.hud.gov/data_sets.html

Strang, Alex Albert Heijn Cuts Costs and Waste for Dutch Shoppers, Canvas8, September 7, 2022. https://www.canvas8.com/library/signals/2022/09/07/albert-heijn-cuts-costs-and-waste-for-dutch-shoppers

Sy, Anne-Gaëlle, ChatGPT kann die Produktivität um 74% steigern, was 51% der Jobs im Marketing abbauen kann, *Sortlist.de*, January 19, 2023, https://www.sortlist.de/datahub/reports/chatgpt/

Ten Facts About World Hunger (n.d.) https://mcusercontent.com/a95cced314f7 6950628adc472/files/a9806e33-0608-e78b-75e1-63ebc1630c2c/Action_ Against_Hunger_10_Facts_About_World_Hunger.pdf

The Impact of the Ukraine War on Food and Agriculture Is Becoming Apparent, ING Report, March 7, 2022, www.think.ing.com

UNEP Food Waste Index Report 2021, UN Environment Program, March 4, 2021 www.unep.org

Utah Housing and Community Development Comprehensive Report on Homelessness (2014). https://jobs.utah.gov/housing/scso/documents/homelessness2014.pdf

Wagner, Patrick Households Waste More Food Than Estimated, Statista, August 20, 2018. https://www.statista.com/chart/15143/percieved-food-waste/

Watson Institute (n.d.) Brown University, *Costs of War Project*, www.watson.brown. edu

Wodecki, Ben, Goldman Sachs: Generative AI Could Replace 300 million jobs, *AI Business*, March 29, 2023. https://aibusiness.com/nlp/goldman-sachs-generative-ai-could-replace-300-million-jobs

Wright, Liam, and Peasgood, Tessa *Cost-effectiveness Analysis of Housing First*, May 2018, https://whatworkswellbeing.org/wp-content/uploads/2020/01/housing-first-Cost-effectiveness-model-may2018_0158328900.pdf

5

PROGRESS IS A PUZZLE

We Need to Get All the Pieces Right

The challenges we have reviewed ensure that we keep our feet firmly planted in reality. Each of these challenges is different and thus requires a tailor-made response. Nevertheless, they share some common barriers to success. This chapter will explore how we can address some of the most persistent and widely shared barriers to lift our chances of success when addressing major, global challenges. These include:

1 Avoiding the familiarity trap
2 History needs to inform our judgements
3 When we shape the present, we are also creating the future
4 Gathering support by engaging others

Avoiding the Familiarity Trap

The pre-pandemic demonstrations by committed, often young, people have put climate change on the agenda. There was media support, and there were at least some enlightened politicians and industry leaders throwing their weight behind these initiatives. However, there is a growing familiarity with what is being done and said, which is likely to blunt the impact of these initiatives. Here are some observations:

A journey of a thousand miles begins with one step

Taking a small step first prepares the ground for taking a more significant step next. For example, dog homes have difficulties getting dogs adopted. A Sydney

DOI: 10.4324/9781003477167-5

dog home developed a programme that allowed office workers to take an orphaned dog for a walk during their lunch break. Once these office workers had taken this small step, they were far more likely to adopt one too.

Kittens are no easier to get adopted. Uber ran a promotion on National Cat Day in February 2015, inviting customers to order a 'Kitten Car' through their app: a kitten from a local animal shelter would be delivered to the customer's door to play for 15 minutes. The promotion was intended as a fun way to promote feline welfare (as well, of course, as promoting Uber). However, those who took the first step by bringing a kitten into their home or workplace were more likely to adopt a kitten. In fact, the programme saw an increase in donations to shelters as well as adoptions. We need to stress that the ultimate goal and the result was positive for the cats – they were not abused by simply being 'loaned out'.

Behavioural science has taught us that nothing motivates people more than success. Each success causes the brain to release dopamine (the 'feel-good neurotransmitter' – see earlier section), and even a small success can make us feel like we have achieved something important. Yet, all too often, the movements organizing demonstrations respond with *'It's not enough!'* when governments or industry promise some action. Of course, they are right: It is not enough. But is this the point?

Alternative approach

We know that people are more likely to take a bigger step once they have taken a small one, that even small successes can be highly motivating, and that it can be hugely effective to take a series of small steps in quick succession, as long as some sort of reward validates those small steps. And, unlikely as it sometimes seems, politicians are people, too!

Rather than rejecting politicians' proposals as inadequate, we should demand small steps that can be taken quickly, praise the decision-makers and actors completing a step without delay, and then demand the next – only marginally bigger – step. Such a strategy is more likely to yield results than a blanket rejection of anything that does not present a total solution.

War and peace

Our brains are hardwired to seek something we want and avoid anything we expect to be harmful or unpleasant. This can make a 'carrot and stick' approach highly effective when trying to shape behaviour.

Consider labour union negotiations: The union makes demands and threatens to strike if these demands are not met. There may be room for negotiating a way forward that is acceptable to both parties, but the point is that the union can apply pressure by *threatening* a strike. The same is true for many other situations: a child might be told that undesirable activities will be punished but rewarded for 'doing the right thing'. An employee can be motivated by the threat of punishment for not delivering but can be rewarded for outstanding performance.[1]

Demonstrations are a highly effective way – possibly the most effective way – to put climate change on the agenda. But repeated demonstrations without any positive feedback mechanism are not likely to drive action, or at least not as likely as a carrot-and-stick approach.

Alternative approach

What if the movement used demonstrations as a threat rather than as a given? Any demands should be for immediate actions that can be validated in the short term so that it will be apparent if the action has – or has not – been taken. The movement should hold off demonstrating if the key decision-makers (typically politicians) promise to take these actions.

When decision-makers act, they earn some praise – and a new demand is made, again for an action they need to commit to and take, with the promise of recognition if delivered and demonstrations if not. If they don't act, a demonstration is triggered.

This carrot-and-stick approach is a way to harness the power of demonstrations to get action rather than just raise awareness. Protests would move from getting the climate emergency onto the agenda to getting some real action!

Familiarity is the enemy of action

We are hardwired to adapt. Humans learned to change their behaviour in response to whatever new challenges our environment presents us with. It's a critical survival mechanism.

Years ago, one of the authors visited South Korea. On the way from the airport, he saw young people congregating on many of the street corners. When he asked his driver to explain, he was told that students met to protest every week, but they had done this for so long that nobody was paying attention anymore.

Familiarity suggests safety, and something that feels familiar is less likely to impact us, to engage us, to make us think or act any differently than we have in the past. This worked well in the hostile natural environment humankind spent most of our history, where familiarity suggested there was no significant threat because we knew the environment and thus could avoid any unpleasant surprises.

In today's world, however, familiarity lulls us into a false sense of security. As we adapt to what is familiar, we change our expectations; as we change our expectations, we start to accept the familiar as a given, if not outright inevitable. Instead of looking for ways to change the situation, we become more likely to shrug and mouth platitudes like 'It is what it is'. We are also constantly bombarded with information, advertisements, and many, many other forms of stimulations and communication. We must filter out information that is irrelevant, meaning that we may filter out the predictable as it offers us little that is new or advantageous.

Familiarity is one of the main reasons we don't deal with catastrophic developments that evolve slowly, such as the climate crisis. We get used to hearing about it, reading about it, talking about it. The ensuing familiarity blunts our feelings, responses, and resolve to deal with it. The problem with staging demonstrations regularly is that, eventually, they become a familiar part of life.

Alternative approach

Demonstrations are highly visible and get a lot of mainstream media exposure – as long as they are news. They are essential to put the issue on the public agenda, but it is arguably more effective to target the key decision-makers once that has happened. More specifically: a politician will typically be more concerned about the next election and the CEO of a major corporation about the financial results that determine their upcoming bonus and continuing tenure. This is human nature.

To get them to act, we need to threaten what is dear to them. This means lobbying voters in their electorate when the next elections come around for politicians. For industry leaders, it means boycotting their companies' products and services and encouraging others to do the same. If that is not an option, it can be highly effective to raise relevant issues at the Annual Shareholders' Meeting where you can raise an issue even if you only own a few shares and generally highlight the company's deficiencies on social media. Fortunately, we see more of such actions in the present day.

Divide, and you will fail to conquer

There are many recent examples of major issues losing traction when the public, politicians, media, and interest groups become polarized. Rational discussions about Brexit were impossible because any statement, analysis, claim, or suggestion made by one party was immediately attacked by the opposing party. In the United States, partisan polarization is now so strong that it has started to impede key functions of the Administration and Congress. These may be extreme examples, but polarization has been a significant barrier to progress worldwide.

Nor does it end with internal issues. An increasing number of nations face the challenge of deciding if they want to cosy up to China or demonstrate 'loyalty' to the United States. Trade wars, military interventions, and wars are creating enemies where we need collaboration on a global scale to address major challenges successfully.

The forming of allegiances is painfully obvious in the issues surrounding the war in Ukraine. Here, the world has sided with Ukraine or Russia and this division is most often based upon the wellbeing of an individual country in terms of the economic cost of supporting either side. Perhaps even more distressing is the role that communication, or rather the restriction of communication, plays in whether a country's population is supportive of one side or the other. Within many countries of the world there is not the freedom to openly access news sources other than those approved by government. Even if you are fortunate enough to be able to go to the websites and television channels of your choice, which of the competing views should you support?

With climate change, the challenge is to bring people on board and get them to contribute to pushing for change, or to take the required action if they are in a position of influence. Staging demonstration after demonstration will eventually polarize the public, politicians, corporations, and the media.

There is evidence that this is already beginning to happen: Police have pointed out that they need a vast force to contain the Extinction Rebellion demonstrations in the UK. The massive associated cost means less money for what the public really needs. Much more problematically, however, they suggest, that their presence elsewhere is thinned down, encouraging criminal behaviour. This lays the foundation for blaming Extinction Rebellion for crimes committed elsewhere due to a lack of police on the ground.

The media – including the 'Letters to the Editor' they publish – has increasingly started to highlight the cost of the demonstrations, including the higher emissions from vehicles stuck at blockaded intersections. The 'flightshame' movement is another example: initially, media reports were largely positive. More recently, however, it has been suggested that it is okay for someone like Greta Thunberg to avoid flying when they are offered the use of a private yacht, but the majority don't have this option.

Alternative approach

All this suggests that it is important to recognize demonstrations as just one tool in the activist's repertoire, and the answer is to use the right tool for the right target at the right time to get the required action. None of this suggests that the public's convenience is more important than the mission to stop carbon emissions, but there is a real risk that the movement will defeat its own purpose by relying on demonstrations as its only tool.

You may disagree with our views. But that's not the point. Rather, we are trying to illustrate that we can learn much from carefully reviewing our actions' effectiveness, taking into account how the human mind works.

History Needs to Inform Our Judgements

Remember the Austrian case example? Understanding history sometimes allows us to understand what drives behaviour, but it can also prevent us from making unfounded judgements.

Let's play a wargame to illustrate the importance of history. Here is a hypothetical scenario: The United States and the Confederate States of America fought the American Civil War. As it happened, the United States won. Let's assume that Hawaii was already part of the States and that the remaining supporters of the Confederate States of America fled to this island, settled there, and declared that Hawaii would be – from then onwards – a separate, independent country.

Assume the role of the President of the United States. What do you believe to be the right action to take? Would you insist that Hawaii continues to be part of the United States and will eventually be reunited with the other states – if necessary, by force – or would you give Hawaii up and recognize it as a separate, autonomous country, governed by the defeated Confederates?

Make your decision, then read on…

Let's look at another scenario: Taiwan was placed under the governance of the Republic of China in 1945. The Chinese Civil War was fought intermittently between 1927 and 1949.[2] In the end, the Chinese Communist Party won and changed China's name from the Republic of China to the People's Republic of China. The Kuomintang, who had led the government of the Republic of China, fled to Taiwan.

Subsequently, both the Kuomintang and the People's Republic of China claimed Taiwan as their own. In 1971, the United Nations expelled the Republic of China and replaced it with the People's Republic of China.

Meanwhile, a democratization process evolved in Taiwan, with the first directly elected president coming into office in 1996. The Democratic Progressive Party (DPP) came to power in 2000 and pursued Taiwanese independence from the People's Republic of China. However, the latter still considers Taiwan an inseparable part of China.

What is the purpose of this brief historical review? The point is that, in our experience, few people know much about Taiwan's history. If they live in the Western world, they tend to accept that China is bullying Taiwan without any acceptable reason for such behaviour. In line with their China containment strategy, the U.S. supports Taiwan and acts as if the People's Republic of China's stated claim on Taiwan is without foundation but rather constitutes an aggressive attack on an autonomous, independent country in the pursuit of territorial gains.

What was your decision regarding the earlier Hawaii scenario? What do you think the United States should have done, faced with this situation? And what is your decision regarding Taiwan? Do you believe China has a claim to this island or not?

To make up your mind about Taiwan's status, you first need to understand the history. That's the solid foundation on which your subsequent value judgements need to rest. Your value judgement will need to balance China's original claim, supported by the United Nations which refers to Taiwan as 'Taiwan, province of China', with Taiwan's impressive efforts to build a country of its own over many decades. How long is the original claim valid? Forever? Or has Taiwan indeed created a new nation, having lived independently for some 70 years?

We should not simply accept whatever we are told by parties that have a political self-interest in a particular outcome. If you only listen to the US-centric media, you are likely to conclude that China is an aggressor that wants to make a land grab. If you listen to the China-centric media, you will likely conclude that Taiwan has been a legitimate part of China since 1945 – and will continue to be so forever.

Today's challenge is that whilst the internet provides us with potentially more information from a variety of perspectives on many contemporary issues, we must learn how to evaluate the sources of this information. This is a challenge as we tend to be too busy to make the effort required to learn about a situation and make up our own minds. It is, of course, easier to simply accept the views of the 'news programme' we favour.

Unfortunately, many organizations that bring you the news also have political affiliations that drive their reporting. What's worse, an ever-increasing number of people get their 'news' from social media! It is hard to believe that they listen to typically ill-informed sources that often have their own agenda. Yet, that's the reality of life. Social media has a lot to answer for and is, in many ways, an uncontrollable whirlwind of destruction to our social and political systems rather than the great democratizing force it was claimed to be when it was first developed.

Our advice: always make sure you understand the historical roots of a challenge, and of the solutions put forward to address it.

When We Shape the Present, We Are Also Creating the Future

Resilience is *the ability to adapt to what we can't change while making positive changes where we can*. This suggests that resilience is a crucial ability or attribute we need to develop and foster during these times of dramatic changes. However, besides adapting and triggering change, there is a third element that is not quite as obvious: The ability to understand the difference. Here, we need to go back to the fact that we are hardwired to give attention to what is immediate and relevant.

Of course, our tendency to focus on what is immediate and relevant favours a short-term judgement of what we can – and can't – change. You may consider any long-term challenges – climate change, the refugee crisis, homelessness, gender inequality, and more – and conclude that there is nothing you can do. You don't have the resources, influence, network, or following to affect change. And you would be right – if you are thinking short-term.

A short-term and personal view was also discovered by Paul in the research we spoke about in the early stages of this book (Hackett, 1995): his research into expressions of willingness to support activities that were designed to help address environmental issues. He found that the features of an activity that were important in influencing commitment included: individuals were aware of the issue, they thought it was important, they believed that organization they were committing to were able to do something about the specific issue and they felt that their personal involvement could make a difference. The research results suggested that we may be responding stronger to situations where we perceive immediate effects, while ignoring stimuli and responses that relate to distant events. However, there was no strong evidence that the short-term perspectives were inevitable.

We may therefore ask the following question: what if you extended your horizon? What if you considered how you could affect change over the next decade or *even over your lifetime*? A change in your time horizon will allow you to see how some positive action you can take will contribute to a better outcome in the long term.

You may not be able to amass any resources even over a more extended period, but you could join others who have the same concerns and want to affect change; you can build your own network, create a following, make your views count using social media, and more. You don't have to give your life over to the cause you want to serve, but even just a little time, spent daily, will make a difference. Importantly, once you start on this journey, you are likely to find that there are plenty of others who share your concerns and, together, you can multiply the influence each individual may be able to exert in their own right.

But the question of time horizon is not only important for each of us who wants to make a difference. It is also essential to consider the time horizon when judging the effectiveness, and even the appropriateness, of actions taken by those in power.

Judging Today's Policies and Actions in Light of Their Future Impact

Some decisions have a long shadow. They are likely to have a determining impact on the sort of world we are creating. Of course, politicians are primarily interested in the short term, like the electoral cycle or keeping *today's* citizens happy. Unfortunately, what may appeal to voters (or citizens in non-democratic countries) may not create a desirable future,

You most likely have come across the argument that the US needs to contain China (Sidra Nasheen, 2021). 'Containment' is essentially a bully's attempt to slow down another party's progress, be this a person, group, community, or country. Of course, the US has powerful means of containment, controlling the international monetary settlement and the US dollar serving as the dominant world currency, with nations and corporations often transacting in dollars and building dollar reserves. The US has also created a dominant, massive military force.

This means the US has the power to slow down China's development. But for how long?

The US's power results mainly from its dominant position after the Second World War. At that time, the US was instrumental in shaping the world's monetary system, international organizations, and trading rules in its favour. On the other hand, China's growth is being driven by an internal growth engine. It is not relying on dominance but has built a huge – and growing – economy based on its large population base, the absolute commitment to progress, and central leadership. The internal growth engine is bound to get stronger and stronger. Containment may slow down China's economic growth, but there is widespread agreement that China will overtake the US as the world's most significant economic power in the long run.

Containment may impress today's voters and build goodwill for today's politicians. It may also help to shore up the market power of American companies. But it will create a future that is steeped in a deeply confrontational relationship. Maybe it does not matter right now, but the future generation of Americans will suffer the consequences of today's containment strategy. They will have to live in a world that is characterized by the invisible forces of cyberattacks, restricted world trade, culture wars, and even major armed confrontations such as the one currently being waged by Russia. This is the next generation's price for the current containment efforts of politicians and power brokers.

Moreover, it will have predictable geopolitical consequences. It will force a closer relationship between Russia, China and North Korea and accelerate the build-up of China's military. It will create a two-tier system in essential developments such as chip technology and will lead to a stronger sense of nationalism rather than supporting the concept of global citizenship.

Whatever we do today is shaping the future. It may not be our future, but it is the future coming generations will inherit. We need to accept responsibility for doing so. If we don't, we will create more intractable, global challenges.

Gathering Support by Engaging Others

Remember Club Recife and their organ donor programme? This worked because their supporters place great value on belonging to their club and will always be ready to support their club's initiatives. Think about the Blue Mountain case, in which the experience of electricity shortages created an immediate and relevant threat that caused the release of cortisol. And the

Iceland case study highlights that we look for dopamine hits, and the best way to shape behaviour is to offer desirable engagement that delivers them. Meanwhile, the German government ensured that miners were looked after when closing coal mines, reducing stress (and thus cortisol production), while creating a feeling of belonging.

In these and other cases, we can observe the power of the mind and how this power can be utilized by triggering the desired response. But how does this work in practice, and how can we exploit this opportunity to shape behaviour?

We have already referred to the hardwired responses and feelings that stem mainly from the time humankind lived in a hostile natural environment. These include belonging, the search for dopamine releases, aggressive competition, the importance of ownership, and more. We all carry within ourselves a host of psychological biases that are largely linked to these hardwired brain circuits.

For example, the *'endowment bias'* suggests that we put a greater value on something we own compared to the same thing we don't possess. Or consider *'social validation'*, which means that we are likely to follow what a large number or majority of people do. But one of the most powerful hardwired responses is the shortcut: Our old brain is designed to take these as urgent action was a key survival factor in a hostile natural environment. Furthermore, given the limitations of the old brain, it could not have accurately analysed a situation in any case. There are dozens of examples of psychological biases that shape behaviour. We need to trigger these biases, which in essence means activating the hardwired responses. These triggers are typically referred to as 'primes'.

Here is a simple way of priming people's behaviour: Creating associations by linking a situation with memories (Roger Dooley, 2012; Stephen J. Genco et al., 2013)

- Barely detectable scents of cleaning products can prime hand washing.
- When a backpack is left in a room people are more co-operative when completing a group task, while the presence of a briefcase makes them more competitive.
- Of a group of people who were interviewing candidates for a job, half were given a cold drink and half a hot drink. The group with the hot drink made more positive comments about the job candidates and would more often make them job offers.

These primes could be explained in a variety of ways: For example, having a warm drink may be linked to memories of relaxing and comfortable experiences, which makes you feel more positive and therefore more receptive to the job candidates you are interviewing. A backpack might trigger memories of student days or camping trips when you were in a group doing things together, putting you into a more cooperative mood.

If you were to become aware that you are feeling cooperative simply because you saw a backpack, you would most likely fight this notion irrational, if not stupid. By doing so, you would minimize the impact of the backpack on your decisions. But when you don't know why you feel cooperative, you let this mood shape your behaviour.

This leads us to an important point: priming works best when people are unaware of it.

We know from studies that primes shape behaviour. They are used extensively by marketers, but have also been adopted by the public sector where several governments around the world have established expert units, often referred to as 'nudge units.' Much of what these nudge units do is to shape behaviour through decision architecture and the use of primes. There are also a number of compelling not-for-profit applications that demonstrate the power of primes.

Here is an example which shows the importance of 'ownership (endowment effect)' and 'social validation.' There are vineyards around the northeast of Vienna. As the city grew, there was concern that the owners of these vineyards would sell their land to developers, and that buildings would replace the vines. There was a lot of support from the locals – especially those living nearby – to keep their world the way it was. But what could they do? Nobody could afford to buy the vineyards. Then somebody had a great idea: why not carve up the vineyards into one square meter blocks, to be sold for a small amount with the proviso that the land would be leased back indefinitely, and at no cost, to the current vineyard owners for as long as the vineyards existed. The proud owners of a square metre would get a Certificate of Ownership that made them feel good and was a great talking point at a dinner party. Without much fanfare, the land was bought, and the vineyards were saved.

Another example of direct community action was experienced by one of the authors. In the town in which Paul lives a developer put in a planning application to develop an area of the shoreline that was much loved by the locals. An online fund-raising campaign was established and instead of fighting the developer the local community purchased the land and dedicated it as a park that could not be built on.

Denmark addressed an even larger issue by allowing for public ownership: The establishment of wind farms. During the 1970s Denmark was a pioneer in developing commercial wind power, something in which it is a world leader today. Entrepreneurship and, most likely, encouragement by the government played a role, but there is another reason for Denmark's success: the cooperative ownership of wind farms.

When built in 2000, the Middelgrunder offshore wind farm was the world's largest. It was co-funded and is co-owned by Danish citizens, who have since recovered their investment and continue to receive a dividend of approximately 7% per annum. The cooperative has 50% of the shares. There is an important cultural factor at play here: Denmark has a long history of cooperative ventures, with cooperatives involved in banking, agricultural businesses, and food shops. Legislation has enshrined cooperative ownership concerning wind farms.

In 2011 the Danish government legislated that new wind farms much be at least 20% community-owned. Experts see the cooperative nature of wind farms as an essential element in aligning the objectives of residents and entrepreneurs wanting to establish a wind farm. One important outcome is that Denmark generates some 50% of its electricity from wind energy and that it is on track to reach 100% renewable energy by 2050.

When you use primes in your communications you may be able to boost its impact significantly. Here are some priming examples:

The Minnesota tax office conducted an experiment, sending four different messages to taxpayers (Richard Thaler and Sunstein, 2008):

1 Taxes go to various good works, including education, police protection, and fire protection
2 Risks of punishment for noncompliance
3 How to get help if they were confused or uncertain about how to fill out their tax form
4 Mentioned that 90% of Minnesotans already complied, in full, with their obligations under the tax law.

Only approach 4 led to a significant increase in tax compliance – a clear case of social validation at work. In the UK, the government has also adopted similar messaging.

The following experiment was staged at an Illinois restaurant (Wansink et al., 2007): Diners were served a fixed-price French dinner, including wine. Half the patrons were told the wine was from North Dakota, and the other half that it was from California. The group that thought they were drinking Californian wine not only rated the wine higher, but also expressed greater appreciation of the food. They also ate 11% more food and were more likely to make a return reservation. The apparent origin of the wine affected diners' perceptions of the restaurant's food, and even the probability that they would return. This is a type of priming which is known as the 'spill-over effect'. Because patrons associated California with great wine, they enjoyed the wine more.

Furthermore, because they enjoyed the wine more, they found themselves in a positive mood, anticipated a good dining experience and, when the food arrived, enjoyed it greatly.

Anchoring can also be a highly effective priming approach. A fund-raising campaign tested three alternative messages (Ross Bernard and Mahmoud, 2018):

1 Would you be prepared to donate to save 50,000 offshore Pacific Coast seabirds from small offshore oil spills?
2 Would you be willing to pay $5 to save 50,000 offshore Pacific Coast seabirds from small offshore oil spills?
3 Would you be willing to pay $400 to save 50,000 offshore Pacific Coast seabirds from small offshore oil spills?

The respective results were: $64, $20, and $143 average donations. The figure mentioned acted as an anchor, i.e., a reference point donors (nonconsciously) compared their donation to. Mentioning a low figure led to lower donations and vice versa.

In another experiment people were asked if they would drive 20 minutes out of their way to secure a $5 discount on a $15 calculator (Thaler and Sunstein, 2008). 68% said yes. However, when they were asked whether they would drive 20 minutes out of their way to secure a $5 discount on a $125 leather jacket, only 29% said yes. The saving is the same in both cases and thus rationally we would expect the same percentage of respondents to be interested in a $5 discount. But the price served as an anchor, making the saving appear less significant in the second case...

Finally, here is a somewhat more complex anchoring application. With a choice of three – increasingly more expensive – options, most people tend to select the middle one: they don't want to 'be cheap', but they want to be smart consumers and not go for the most expensive option. In other words, the two extreme options work as anchors, creating a comfortable place for the middle option. This approach can be used for any investment. For example, you may want volunteers to spend some time helping. By showing three options, ranging from a highly time-consuming activity to one that is almost laughable because it hardly takes up any time at all, you may find that most people select the middle option.

Sometimes it is sufficient to set the agenda, which can be achieved by simply triggering brain circuits. Here is an example: Half of the customers looking to buy a computer in a store were asked about their memory needs and the other half about their processor-speed needs. The group that was asked about their memory needs ended up buying computers with larger memory, and those in the other group ended up buying computers with higher processor speeds. Getting them to think about specific attributes of the product affected their decision in favour of that attribute.

We have already referred to the 'endowment effect'. Here is a real-world example on how to trigger the 'ownership' brain circuit: Half the customers of a German car wash service were given a coupon that required them to pay for eight services to get one free. The other half was given a coupon requiring ten services to get a free service, but with two already marked.

Both versions required eight paid services to get one free service. However, the redemption rate went from 19% to 34% when the second version was used, representing an uplift of 79%. The reason: the two stamps created a situation where customers felt they already 'owned' two stamps which they didn't want to give up – so they went on to collect the remaining eight.

Being asked to make an effort can create friction, which is often a barrier to action. For example, reducing the number of fields a donor must fill in as part of a registration from 11 to 4 led to completed donations increasing by 140% (Ross Bernard and Mahmoud, 2018). It is always nice to get more information when supporters register or take up a membership. But balance your need for information with the person's desire to go through a frictionless process.

An incremental approach may allow you to create the momentum you would like to generate. Here is an illustration (Robert Cialdini, 1984): Two offers were tested with different samples of respondents:

Group 1 was asked: 'Would you be prepared to volunteer to spend a day at the zoo as a companion for individuals with a mental health challenge?' 17% said yes.

This was followed by 'Would you consider committing to being a two-year counsellor for the same group of individuals (with a mental health challenge)?' 100% said no.

Group 2 was asked the same questions but in reverse sequence. In this case again 100% said No to the two-year commitment, but 37% said Yes to spending a day.

When your message gives people the feeling that they are making an effective contribution they are more likely to support you. For example, they may feel that they can't give enough time or money to make an impact. Pre-empting this barrier to action can deliver better outcomes (Ross Bernard and Mahmoud, 2018). Here is an illustration: The American Cancer Society found that 28% of people gave when asked 'Would you be willing to help by giving a donation?' When the invitation was changed to 'Would you be willing to help by giving a donation? Every penny will help!' almost twice as many donations were received!

Another barrier to supporting a cause or organization is the feeling that the contribution will not lead to desirable action. Perhaps it will be used for administrative expenses, or promotions – who knows? Again, it is a matter of preempting this concern. When Doctors Without Borders changed 'We need your gift to respond to this crisis' when approaching regular donors to 'Thanks to your regular gift our team has already left for the crisis zone' they got a 16% increase in response rates.

Similarly, sensory experiences can also trigger memories. Smell, visual images, sounds, haptic sensations (touch) and taste can all trigger memories that create certain moods or encourage specific behaviours. This is why, for example, the smell of bread is not only used by bakeries and food stores, but also by real estate agents to trigger thoughts of a homely environment with fresh bread being shared by the family. The smell of chocolate triggers a feeling of wellbeing in many people. It has been shown to work even in environments where chocolate is not available. For example, in a bookshop it resulted in customers spending more time in the store and buying more books. If you have a place where you sell items, like a second-hand store, or a place where you recruit volunteers or where they meet, you can use scents as a means of creating positive moods and a feeling of belonging.

Don't assume that priming is only effective with an ignorant audience. Here is a compelling experiment (Daniel Kahneman, 2011): German Judges, each with an average of more than 15 years of experience in courtroom decisions, read a case about a woman who has been caught shoplifting. They were then asked to roll a pair of dice. The dice were loaded and always showed either a 3 or a 9. When judged rolled a 9, they suggested on average an 8-month prison term. When they rolled 3, they suggested, on average, 5-month prison terms.

This completes our brief exploration of tools that can be used to boost the effectiveness of behaviour change programmes. We are almost at the end of our journey. But there are a couple of important points left to consider before we reach the final page. More specifically, we need to explore what you could do to get yourself battle-ready!

Notes

1 It should be noted, however, that reward is a much better way of bringing about change in a desired direction. First, rewards are likely to bring about longer-lasting changes as long as the desired action is still rewarding. Rewards are also perhaps more likely than punishments to foster a positive belief and attitude system towards the behaviour that is being rewarded. Punishment, on the other hand, often results in people developing ways to avoid being caught out and thus to avoid punishment rather than adopting the desired behaviours. However, both reward and punishment can be used in tandem to bring about change.
2 Wikipedia provides an outline of Taiwan's history.

References

Cialdini, Robert B. (1984) *Influence: The Psychology of Persuasion*, New York, United States: Harper Business.
Dooley, Roger (2012) Numerous examples can be found in: *Brainfluence. 100 Ways to Persuade and Convince Consumers with Neuromarketing*, New Jersey, United States: Wiley, Hoboken.

Genco, Stephen J., Pohlman, Andrew P., and Steidl, Peter (2013). *Neuromarketing for Dummies*, Hoboken, New Jersey, United States: For Dummies (Wiley).

Hackett, P. M. W. (1995) *Conservation and the Consumer: Understanding Environmental Concern*, Milton Park, Abingdon, Oxfordshire, UK: Routledge.

Kahneman, Daniel (2011) *Thinking, Fast and Slow*. New York, United States: Farrar, Straus and Giroux.

Nasheen, Sidra, US Policy of Containment Against China, *Paradigmshift*, March 17, 2021. https://www.paradigmshift.com.pk/us-policy-of-containment-against-china/

Ross, Bernard, and Mahmoud, Omar (2018) *Change for Good. Using Behavioural Economics for a Better World*, London, UK: The Management Centre.

Thaler, Richard, and Sunstein, Cass R. (2008) *Nudge. Improving Decisions About Health, Wealth, and Happiness*. London, UK: Penguin Books.

Wansink, B., Payne, C. R., and North, J. (April 23, 2007) Fine as North Dakota Wine: Sensory Expectations and the Intake of Companion Foods. *Physiology & Behaviour*, 90(5), 712–716.

6
GETTING READY TO ACT

The traditional approach to addressing a challenge is to search for a solution and take appropriate action. Boom! It's done. That's great – but only if the problem does not mutate like a virus.

The traditional approach works well in a reasonably stable environment. Changes tend to occur incrementally and over a more extended period. Sometimes there is an interruption – say a recession – but the world returns to 'normality' soon after. In such situations, we can analyse a problem, work out the underlying factors causing it, and act to effectively address our challenge.

However, we now live in a rapidly changing world and, importantly, problems are often interconnected. Take, for example, the way governments of leading nations dealt with the Covid-19 pandemic. They pumped a lot of money into the system to support businesses and individuals. A laudable effort but, unfortunately, this caused inflation and led to a significant widening of the economic inequality gap.

The pandemic also exacerbated other problems. It led to an increase in homelessness, mental health issues, drug use, and so forth. It played into political divisions, boosted 'fake news', led to violence, and widened the gap between wealthy and poor nations. The support provided by governments led to massive fraud. By August 2023, the US Justice Department announced that they had seized $1.4 billion in Covid-19 relief funds that criminals had stolen and that they had charged 3,000 defendants with crimes (US Department of Justice, 2023). The department also announced the launch of two new Covid-19 Fraud Strike Forces, suggesting that this may be only the tip of the iceberg.

DOI: 10.4324/9781003477167-6

You may think of the pandemic as a 'once-in-a-lifetime' event, maybe in the same way as many people believed the First World War to be – except this turned out to be wishful thinking!

We are facing a multitude of massive, complex challenges that are often interrelated. Our solutions may have unintended consequences. And, importantly, these challenges are not static, but evolve and morph in often-unpredictable ways,

Here are a few examples: Climate change is a central issue because it (potentially) impacts many other challenges. Climate change is expected to have a massive impact on food security, water shortages, the number of refugees and homeless, and some experts believe it will even facilitate future pandemics as changes in weather favour particular species that may carry transmittable pathogens.

The pandemic is another central challenge because it diverts attention and budgets from other challenges due to its imminent danger to everybody. We can also expect it to have had a disproportional impact on a number of disadvantaged groups: refugees who will be forced into overpopulated camps with little or no vaccinations made available to them; the homeless who will also fall through the cracks when it comes to vaccinations; and malnourished children and adults due to their low levels of resistance and pre-existing conditions.

Arguably, however, the most severe barrier to progress is self-interest and the emphasis on short-term personal gain. After a long period of economic growth that created a stable environment, we are now looking at the need for a multitude of transitions. From the fossil fuel industry to the military and the associated military-industrial complex, to workers in essential industries that will need to introduce a much higher degree of automation. Somehow, we need to balance patent laws protecting companies and the need for access to pharmaceuticals and other essential goods in times of catastrophic challenges. We need to change from fossil fuels to renewable energy sources and from water-intensive food production to foods that can feed the world without worsening water shortages. If we want to succeed, we also need to put a lid on partisanship, recognizing that there is no point in winning elections when there is only chaos to be governed. Importantly, we need to shift our leadership style from combative to collaborative, something which many of today's leaders are not well equipped to do.

Most likely you will agree with us when we say that this will not happen. There are too many vested interests, too many entrenched attitudes and values, too much for too many powerful parties to lose. It is also important to note that we are not suggesting that we have the answers to all the world's problems. Many of the issues are intractable and ones for which we have no clear solution. Population growth is one such problem which, if it continues at anything like its current pace, will eventually overwhelm our available resources. Of course, there may be natural disasters, plagues, floods, and the like and also self-inflicted human atrocities, such as wars, that reduce our numbers. However, these are hardly factors that we can build into models for our sustained future.

Yet, coming back to the Curve of Doom, we do see a time when this will happen: When we have reached a collective sense of deep despair that is much stronger than self-interest, ignorance, money, and power. It is unlikely that this will happen unless we first go through a period of disasters that make the current pandemic and climate change impact feel like a walk in the park. But given the road we travel on – and as we are not prepared to leave it – it will happen eventually.

You are wrong if you think we are resigned to accepting this fate. But it is crucial to be realistic. It is important to adapt to what we can't change, but it is equally important to make changes where we can. And we can still do so, and our actions can make a difference. At the time of writing, more than 6 million people have lost their life, defeated by the pandemic, and experts believe the actual figure is more likely between 15 and 20 million. The United Nations reports that hunger kills over 7.6 million people annually – amounting to a daily figure of around 21,000 people – and children count for a disproportionate share of these deaths. Similarly, the number of refugees dying while trying to reach safety is in the thousands, and nobody knows how many die in camps or somewhere on the road. But we do know that recent wars have killed tens of thousands, including many civilians killed as collateral in bombings and drone strikes. The war waged by Russia in the Ukraine and by Israel in the Gaza Strip has, at the time of writing, claimed the lives of thousands and caused truly enormous infrastructure damage.

What if we could, collectively, make a dent in these statistics? Even a small dent would mean thousands of lives saved and tens of thousands being able to see a worthwhile life ahead of them, rather than just a picture of hopelessness and despair. It is not the nature of our challenges that prevents us from addressing them. It is us. The natural Commitment Gap we need to overcome.

You are exactly the person we want to talk to if you are willing to make whatever contribution you can offer. We are not trying to tell you what to do to get battle-ready, but some of our suggestions will perhaps be useful.

Developing a Thinking Habit

Helpful, professional people previewed this manuscript and suggested not to include this chapter. It was seen as too demanding, too abstract, requiring an effort many readers would not be prepared to make. But, as you can see, we did not heed their advice. The reason is our absolute conviction that many people want to understand the tools and processes that can help us address major global challenges. It will be up to you to look at the evidence and make your own judgement. If you are not interested, there is a simple solution: Just skip the rest of this chapter.

If you are still reading, let's review some tools that may help us make sense of the major challenges we are facing and develop ways of addressing them. Our consideration of tools should be read with the notion of the 3-stage model for commitment to civic and environmental action: knowledge; importance; and effectiveness.

Question 1: What Do We Know?

This is an obvious starting point. Here, the source of information is important. Obviously, when you rely on some social media post, there is every chance that the writer does not know much and, worse, that they may have their own agenda, i.e., trying to persuade you to adopt their views and ideas rather than help you make up your own mind.

Whenever possible, you should rely on sources with a sound reputation and a desire to uphold it. Avoid sources that have their own agenda, whether political, commercial, conspiratorial, or otherwise. But even when you are dealing with reputable sources, you need to always use your own judgement and, where possible, corroborate the findings by looking at several sources.

Of course, there is also a real danger of getting overwhelmed. There are numerous parties writing papers and reports, staging conferences, establishing interest groups, conducting interviews, providing comments and suggestions. An information overload may cause you to give up, deciding that there is no way you can work your way through the issues. Don't! Be selective and focus on just a few reliable sources. Anything you learn from your exploration will put you in a stronger position.

Experts

Whenever possible, you should consult experts. I am not talking about paid consultants here, but people who have explored the challenge you are addressing or related areas. They are often aware of what has been tried before and how it worked out. They may also be aware of issues, barriers, and opportunities you have not considered. Again, you need to avoid experts with a personal interest in the matter as they will use their position to try to influence you, rather than provide you with useful information that helps you make up your own mind.

What are their experience and expertise? Check them out on LinkedIn or just do an internet search. Have they written something, posted something? What are they experts in? Don't trust an executive's advice on health or a doctor's advice on the economy. I am not saying their views should not be considered if you find them compelling. The point is you should check the underlying assertions and assumptions before you accept their views.

Don't ever think that experts won't be happy to help. They are often more than helpful and can save you untold hours pursuing dead ends. For example, when we worked on the Australian Blue Mountain case study, we consulted Prof. Siobhan McHugh, an expert in this field, and she clarified many aspects about which we were unsure.

Survey Data

Sometimes there is available survey data, showing what other people think. Again, this can be useful, but you need to remember again that survey respondents may be biased, may be unsure of their answers, or may have their own agenda. Above all, ask: Where did the sample of respondents come from? And, who are they representative of?

For example, we know that many respondents claim they will become organ donors; prefer to buy goods from ethical corporations; plan to get vaccinated; always recycle; are committed to supporting the homeless, and so forth. However, observation studies and data on actual behaviour often tell us that these answers do not reflect reality. In a particularly informative study, researchers interviewed grocery shoppers when they left a supermarket, asking them if they bought healthy food and favoured products manufactured in their own country. They then checked the shoppers' products in their trolley and found a big gap between what they said and what they did!

But the problem does not end there. Some respondents will not be certain of their answers. They will hesitate and think about how to respond, but once they give an answer it will carry the same weight as an answer given with no hesitation because the respondent was sure about their views.

Finally, respondents may want to present themselves in a particular way – as progressive, concerned, high-achievers, thought-leaders, representing the views of a political party or religion, et cetera.

As you can imagine, this leads to survey results that are less than reliable, especially when they cover sensitive issues or issues respondents have never before thought about. Unfortunately, the major global challenges we are interested in tend to be sensitive, and surveys are likely to address issues respondents have never considered before. So again, tread carefully.

On the positive side, survey data may allow you to see the difference between countries, age groups, gender, and more. It may offer you a historical perspective, showing how attitudes and opinions have changed over time. We are not suggesting that you should ignore survey data. We are simply voicing a warning: be careful and consider possible biases that may have affected the results.

You may at this point reflect upon the fact that questioning someone about their opinions, behaviours and the like is inherently unreliable – and you'd be right. As we are trying to understand and predict people this puts us in an awkward position. We may therefore decide that instead of asking people about their behaviours and receiving edited replies, we watch their actual behaviour. This, you may reason, must be reliable as we are watching someone actually doing something.

This is true. If we ask someone if they will switch their coffee purchases to sustainably and ethically sourced coffee, their responses will only be a partial reflection of what they actually do. Unfortunately, observing someone actually buy ethically sourced coffee (for example) also has inherent problems as we are unable to know why they made this purchase. Did they buy the coffee because of the awareness-raising advertising campaign? Or perhaps they have made the purchase because they usually buy tea, which was not available on that occasion. Alternatively, they may have bought the ethically sourced coffee because it was cheaper, because they were with friends who support ethically sourcing, and so on. In all these cases, we are unable to understand the key drivers of their behaviour.

As it seems that all forms of information have their weaknesses or flaws it is probably a good idea to try to gather information that has come out of a variety of different types of research and has a variety of trustworthy sources.

Question 2: What Does It Mean?

In dealing with our second question, we have found Force-Field Analysis, developed by Kurt Levin, useful (Mark Connelly, 2020). Challenges and opportunities do not exist in a vacuum. They occur because of several, often-contradictory, factors determining a particular outcome.

Think about something you want to do or achieve. For example, a job you want to get, a skill you want to learn, the weight you want to lose. There are invariably some positive factors that help you to make the change. It may be easier to get the job you want because of your relevant skills, reputation, a network of contacts, or ability to be persistent. But there are invariably also some negative factors that will make it difficult to get the outcome you want – for example, intense competition for the job, lack of experience, or lack of self-confidence. In fact, if there were no negative factors you would already have the job in question!

The balance between positive and negative factors will determine the outcome (in the example we used whether or not you will be successful in securing the new job). It follows that you need to change this balance in your favour to increase your chances of getting the outcome you want.

You are more likely to get the desired outcome if you strengthen positive and/or weaken negative forces. When you do this, you change the balance of

**We want refugees to get
a new life**

Positive Drivers
>>>>

<<<< Negative Drivers

It's the 'right thing to do'
Refugees can help us address one
of our challenges

Fear of:
• don't like change
• refugees are strange people
• we will be losing jobs to underpaid refugees
• refugees bring violence, terrorism with them
• we will be losing our identity
• added government expenses will lead to higher
  taxes

I had to work to get accepted as an immigrant –
they shouldn't have it easier

FIGURE 6.1   The Force-Field concept.

the Force-Field in your favour. But even if you can't find a way of shaping the forces at work, you will gain a much better understanding of the challenge you are facing, and you may gain a new perspective that will help you develop an effective, winning strategy.

Figure 6.1 offers an illustration.

Before we move on, here is a brief overview of the process undertaken when conducting a Force-Field Analysis:

1 Before you start, very clearly define the outcome you want to achieve.
2 Draw up a Force-Field template. On the top of it, write the outcome you would like to achieve.
3 On the left side, list all the positive factors that may help you to achieve the outcome you desire. On the right side, list all the negative factors that are barriers to success.
4 When you have finished listing the positive and negative factors, go through each of the two lists and select the most important negative factors (i.e., those that constitute the most significant barriers to achieving your desired outcome), and the most important positive factors (i.e., those you can build on to achieve the desired outcome).
5 Once you have selected the most important positive and negative factors (i.e., those factors you believe have the greatest impact on the outcome), search for ways of weakening negative or strengthening positive factors. Alternatively, you may consider creating additional positive factors that would change the balance in favour of the outcome you are seeking.
6 It is typically easier to deal with the negative factors first. Just start with the most promising negative factor first, then move on to the next one, and so on.

7 When you have covered the most important negative factors, move on to the positive ones. With the positive factors, your objective is to find ways of strengthening or building on these factors.

8 When you run out of ideas, take a break, and then review the results of your session. It is good practice to put your analyses aside for two or three days and then review them. There are two reasons for this: first, you gain some distance by not actively pushing yourself to find solutions. Second, your nonconscious mind will turn the challenge over without you even being aware that this is happening. Therefore, people often have the best ideas and insights when they are not actively trying!

9 Select the idea that has most chance of success and develop it into a plan of action.

Question 3: Could I Have Overlooked Something Important?

Scientist, researcher, and corporate advisor Gary Klein put forward an interesting concept (Gary Klein, n.d.): The pre-mortem.

We are all familiar with the post-mortem. Once a significant milestone has been reached or a project completed, we analyse what happened. What worked, what didn't work, and why? These reviews are hugely important as we learn by experience, and unless we take time to explore and understand what we have experienced, we learn very little.

However, Klein wondered if it would make sense to also conduct a review *before* embarking on a project or program. In this pre-mortem, you would essentially imagine that you have taken your carefully considered action, and the results have been catastrophic: total and utter failure. The worst possible outcome has materialized. And you are asking yourself: *'How could this have happened?'*

In other words, you are pretending your actions or plans have failed and then exploring what could have happened to get such an unfavourable result. The benefit of this pre-mortem is that it sensitizes you to what *might* lead to failure, allowing you to identify any weaknesses in your approach *before* you have taken any action.

We have found the pre-mortem a truly useful tool, and it is very simple to apply it. You imagine that your actions or plans resulted in total failure and ask yourself *'What could have gone wrong to lead to such a disaster?'* You simply write a list of what could have gone wrong. Don't include minor details – things that might have caused a little setback. Instead, you need to focus on what could have led to catastrophic failure.

Keep your list 'open' for a couple of days and add to it. Then review it, eliminating duplications, and relatively minor factors. On completion of your pre-mortem, you may reach one of the following conclusions:

1 Your plan of action is sound – you could not identify anything that could result in a catastrophic outcome (this is very rarely the case, and you may want to review your assessment!).
2 There are elements of your action plan that need to be reworked as they could cause catastrophic failure.
3 There are elements of your action plan that could cause catastrophic failure, but you cannot address them at this stage. However, being now aware of these risks, you can monitor them and adapt your actions to address them should they start to cause serious problems.

But you can also use the pre-mortem to assess solutions put forward by others, like your government or a self-appointed expert. Let's take the Technological Revolution example we introduced earlier to demonstrate this. The pre-mortem has identified the following factors that could create a catastrophic result:

1 Progressive Sector companies are reluctant to pass productivity gains on to the consumer unless forced by competition (i.e., some companies passing the savings on to gain a price advantage).
   <u>Solution</u>: We may need stricter consumer protection laws and close monitoring of monopolies and duopolies as they are not exposed to competition, or only in a very limited way.
2 We have cut work hours too much, leading to a negative impact on the economy.
   <u>Solution</u>: The ability to introduce shorter working hours in small increments reduces the risk of over- reacting. If the Technological Revolution's impact is less than expected, it simply means that work hours won't be reduced much, or only after long periods of stability. If the impact is greater than expected, it will lead to a more dramatic and faster reduction in work hours. Importantly, this approach also allows each country to adjust work hours based on the local impact of the technological revolution and its own balance between progressive and stagnant industry sectors.
3 There is resistance by many industry leaders and investors who see this approach as limiting the freedom of enterprise. Let's not forget that the intent is to restrict the ability of companies in the Progressive Sector to capture the additional margins resulting from technologies replacing workers. Naturally, those managing these corporations and their investors would love to enjoy these massive margins rather than reducing work hours while keeping workers' pay at the same level.
   <u>Solution</u>: We need to raise consumer awareness of the benefits and publicly shame companies that do not pass on benefits. This may sound like a pipedream, but we have seen this work with sustainability. As soon as consumers started to express a preference for companies that could claim sustainable operations, we have seen more and more companies taking action to claim that status.

4   There will be a requirement for extensive retraining as workers from Progressive Sector organizations will need to find jobs in Stagnant Sector ones.

   <u>Solution</u>: Here, technology can help by facilitating the delivery of training programs online at a low cost.

As you can see, conducting a pre-mortem delivers an understanding of what the risks are, allowing us to find a solution before the problem occurs or, at least, to develop a Plan B. This allows you to deal with weaknesses in your plan before you take any action. It is also a great approach to assessing how effective solutions put forward by others might be.

References

Connelly, Mark *Force Field Analysis – Kurt Lewin*, September 12, 2020 www.change-management-coach.com
Gary Klein (n.d.) For pre-mortem method of risk assessment visit www.gary-klein.com
Justice Department Announces Results of Nationwide COVID-19 Fraud Enforcement Action, Office of Public Affairs, US Department of Justice, August 23, 2023. https://www.justice.gov/opa/pr/justice-department-announces-results-nationwide-covid-19-fraud-enforcement-action

7

ACTION! WHAT CAN YOU DO?
WHAT CAN OTHERS DO?

When we conducted a large-scale survey in Germany in 2023, we asked 1,000 respondents what they were seeking in these challenging times. We used the Response Time Test to gather not only explicit, but also implicit data (Peter Steidl, 2023).

Cognitive science research has shown that a strong connection between neuronal patterns leads to an immediate response, while a weak connection causes hesitation. In other words, when a respondent is convinced of their response they will answer without delay, whereas if they are unsure or base their answers on pretense, they will hesitate. The Response Time Test measures calibrated reaction time to determine how strongly respondents are convinced of their answer. The value of being able to distinguish between those who strongly believe in their answers versus others who are uncertain becomes apparent when we start to look at the survey findings.

You may recall the point we made earlier about respondents often delivering the answer they feel is the expected or 'right' one. Who wants to say they are not concerned about climate change, don't buy brands committed to sustainability, or feel that homelessness is not a high-rating concern of theirs. Using a research methodology that delivers not only explicit, but also implicit data, identifies respondents who are committed to their answer as well as others who are not (see Figure 7.1).

Here is the legend, telling you how to 'read' the results. In this example, 70% responded with 'yes' and 30% with 'no'. These are the explicit answers. 60% are uncertain or pretend by giving what they consider an expected answer or one that positions them the way they want to be seen. This group is made up of 40% who tend towards a positive and 20% who are closer to a

DOI: 10.4324/9781003477167-7

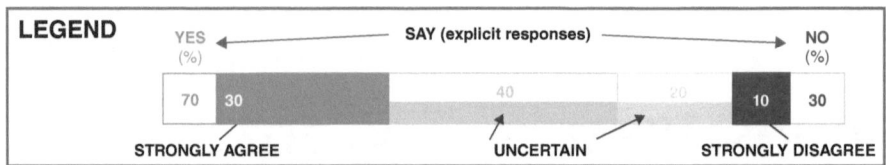

FIGURE 7.1   Response Time Test Legend.

negative answer. When you try to influence behavior, it is the group of uncertain respondents that represents your target group. The committed 'yes' segment is already on board and the committed 'no sayers' are not likely to change their mind. Thus, the uncertain segment is the territory where you can win more supporters.

The survey results delivered the following explicit and implicit data (see Figure 7.2).

Most of all, people want leadership, i.e., someone who will lead them through this time of crisis, or someone they can support because they are doing something positive. Next desired is the opportunity to just withdraw and escape being forced to think about the major challenges we are facing. This is followed by a desire to belong, to get the feeling of greater security that comes from joining with others in difficult times. And, finally, with only a small percentage agreeing, they want someone to tell them how they themselves could contribute.

The latter results are disheartening. Short-term escape – perhaps playing a video game or enjoying a holiday – allows us to recharge and get battle-ready. But a longer-term withdrawal means that we are not contributing to addressing the challenges that impact negatively on our quality of life. Indeed, individuals' desire to seek the solution elsewhere rather than to contribute actively themselves is the biggest barrier to progress.

In this section we will explore how various parties can contribute to addressing some of the challenges we are facing. At the same time, we will be emphasizing the important role individuals have to play, even if they don't have resources, they can deploy to address challenges directly.

Industry & Commerce

There are two questions we need to address: First, what is Industry & Commerce doing to contribute to addressing some of today's pressing global challenges? And second, what is their motivation – why do they engage?

Let's start with the first question. There is ample evidence that there are companies contributing to addressing major challenges, and some even take a leadership position. Here are some examples.

What you want to see in this time of crisis?

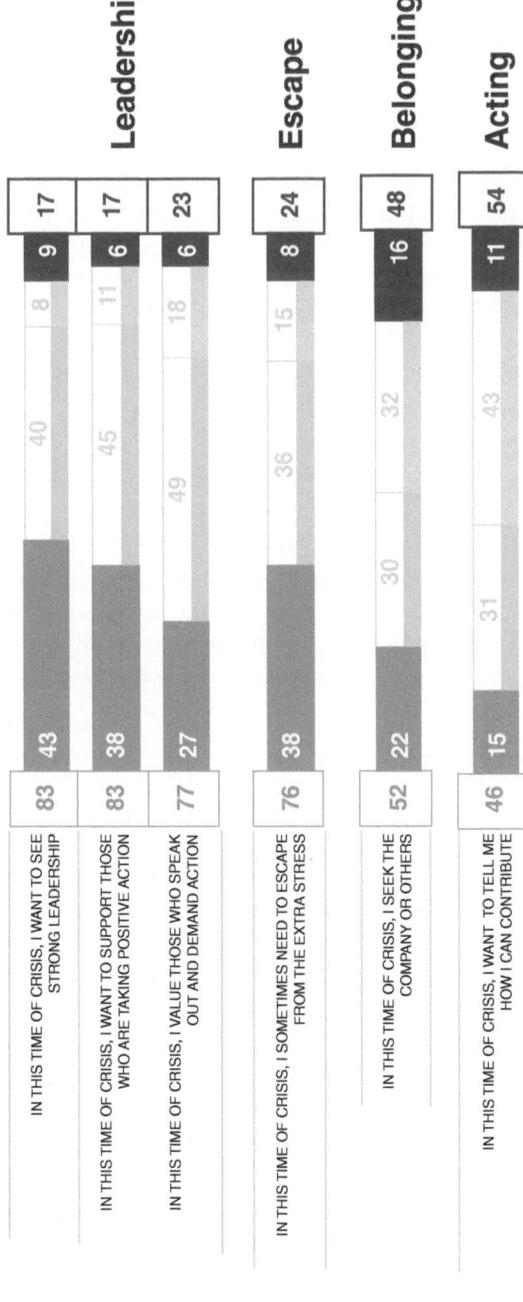

IN THIS TIME OF CRISIS, I WANT TO SEE STRONG LEADERSHIP	83	43	40	8	9	17	**Leadership**
IN THIS TIME OF CRISIS, I WANT TO SUPPORT THOSE WHO ARE TAKING POSITIVE ACTION	83	38	45	11	6	17	
IN THIS TIME OF CRISIS, I VALUE THOSE WHO SPEAK OUT AND DEMAND ACTION	77	27	49	18	6	23	
IN THIS TIME OF CRISIS, I SOMETIMES NEED TO ESCAPE FROM THE EXTRA STRESS	76	38	36	15	8	24	**Escape**
IN THIS TIME OF CRISIS, I SEEK THE COMPANY OR OTHERS	52	22	30	32	16	48	**Belonging**
IN THIS TIME OF CRISIS, I WANT TO TELL ME HOW I CAN CONTRIBUTE	46	15	31	43	11	54	**Acting**

FIGURE 7.2   Illustration of survey results using the Response Time Test.

Addressing the Refugee Crisis

Russia invading Ukraine has caused yet another massive flow of refugees trying to escape the horrors of this war. Quite a few companies are contributing to making it a bit easier for refugees to deal with their challenging situation:

- Because of its vicinity, Poland attracted the largest inflow of refugees. Mastercard and McCann Poland created the Where to Settle app to help the 1.5 million Ukrainian refugees in Poland to choose the best location within Poland to live. By July 2023, the app had helped some 300,000 refugees (Raini Kenyon, 2023a, 2023b).
- Airbnb provided short-term housing to 100,000 Ukrainian refugees (Nicole Bosky, 2022).
- Eurostar offered complementary travel for eligible Ukrainian refugees who took a train in Paris, Brussels, Amsterdam, or Lille to go to London (Luana Sambell, 2022).
- Some of the world's biggest brands formed the TENT Partnership for Refugees to train 250,000 refugees across Europe and offer more than 13,000 jobs (Bruno Davey, 2023).
- Action was not limited to major corporations. For example, software developer Chili Piper raised more than £170,000 to support NGOs active in Ukraine and built Bridge, a digital platform that helps aid agencies to connect with organizations offering resources (MaryLou Costa, 2022).

Contributing to the Circular Economy

Plenty of companies are playing their role in creating a Circular Economy:

- Selfridges' Worn Again Season Series launched a collaboration with Loanhood, allowing shoppers to trade in unwanted clothing for credits to spend on other secondhand goods in-store. The retailer plans to have 45% of its transactions aligned with the Circular Economy by 2030 (Nicolas Lopez, 2023a, 2023b).
- The Levi's Buy Better Wear Longer campaign encourages consumers to re-wear and repurpose their clothes. The company is also leading research and development to create sustainably produced denim that drastically reduces its water footprint (Josh Greenblatt, 2022a, 2022b).
- Twig is an app enabling users to cash in on their existing goods. Users simply upload an item to Twig, which assesses its value algorithmically by sifting through real-time selling prices across multiple platforms and makes an offer. When accepted, the user receives immediate payment before even shipping the item (Eva Clifford, 2022a, 2022b).

- Murfy offers easy and efficient repairs to home appliances that people would normally replace unnecessarily. Customers book a time slot and pay a fixed fee, and within 48 hours a technician arrives (Riani Kenyon, 2023a, 2023b).

Facilitating Small Actions That Can Make a Real Difference

Many companies also recognize that individuals would like to play a part in addressing some of the challenges we are facing and, to that end, facilitate their involvement:

- Tesco opened the Give Back Express in central London in collaboration with FareShare and the Trussell Trust, encouraging people to donate the 25 most-needed grocery products as identified by the charities. Tesco's 'take stock' initiative helps people reduce food waste to save cash and limit their environmental impact (Raini Kenyon, 2022).
- Co-op grocery stores prioritized a community food hub over funding a 2022 Christmas advert (Raini Kenyon, 2022).
- Burger King France set out to help local potato farmers by buying 200 tonnes of potatoes. In February 2021, the fast-food chain handed out two-pound sacks to every customer going through its drive-through outlets, together with a message encouraging customers to make 'a resolution for 2021: to keep buying potatoes'. For many consumers, this was not a tall order – they would have bought potatoes anyway, but now they could feel good about it, and of course feel good about Burger King (Leila Zadeh, 2021).
- Cheil Worldwide launched the Seanack campaign in Korea in 2022, in response to the anticipated increase in littering of the shoreline as beaches were opened for the first time in three years post-pandemic. Beachgoers take litter they have picked up along the beach to a Seanack station, where they can exchange it for snacks (Avinash Akha, 2022).

This leads us to the second question we raised:

Why Are Companies Engaging – What Is Their Motivation?

We are not suggesting that the key decision-makers in these organizations don't have personal values that encourage them to use the resources of their respective organizations for good. But they are also responsible to shareholders and their fiduciary duty is to generate a return to shareholders. In other words, we must consider their actions in the context of shareholder return. The commercial value of their initiatives depends almost totally on how their actions resonate with their customers and potential customers.

Of course, there are also other target groups, such as shareholders, regulatory bodies, business partners, and the media. For example, when attempting to expand a business into a new market, investing in some local not-for-profit project can make it easier to get support from local regulators. Similarly, it may be advantageous to support an initiative in which a potential business partner already has a stake. And we do need to also acknowledge that 'doing good' can help a company gain positive media exposure.

However, the main objective is typically to strengthen the loyalty of current customers and attract new ones. Here is some data supporting this proposition:

- Europeans are more likely to buy from a company that hires refugees (51%) and advocates for the government to accept more refugees (39%) (Raini Kenyon, 2022).
- A survey by Harris Poll revealed that 82% of people want a brand's values to align with their own, and that three-quarters of people have stopped buying a brand over perceived differences in such values (Bruno Davey, 2023).
- 60% of Gen Yers are inspired to have more sustainable wardrobe habits after experiencing multiple lockdowns (Nicolas Lopez, 2023a, 2023b).
- Globally, 72% of consumers are looking to adopt circular practices. 81% of Britons prefer to purchase products from sustainable sellers (Eva Clifford, 2022a, 2022b).
- 76% of Gen Zers and 79% of Gen Yers expect brands and retailers to become more sustainable, and 75% of Gen Zers are most likely to shop with brands that align with their personal, social, or environmental values (Josh Greenblatt, 2022a, 2022b).

Imagine if nobody took notice of the actions companies take to help address some of the major global challenges. They would be much less likely to invest some of their funds and employee time in initiatives of this nature. In other words, the main driving force is individual consumers, who have much more power than they may realize. Individuals can encourage companies to do good by rewarding them, that is, buying their products or services, recognizing their contribution on social media, recommending them to others, and so forth. This is an option open to everyone who is in the market for the types or products or services a company offers.

Media

Most media organizations are businesses, but we felt we should cover the media separately because of its unique influence on people's knowledge, attitudes, opinions, and beliefs. Furthermore, the issues are somewhat different.

The media should help citizens understand important issues, investigate, and highlight illegal or unethical conduct, and provide encouragement and support when it comes to addressing the major challenges we face today. Unfortunately, many media organizations are not playing this role, but rather lend their voice to subversive elements or try to boost their revenue without regard to the reliability or quality of reporting. There is also the issue that even the most responsible media outlets need to maintain reader/viewer interest, and often choose to feature stories about topics that they know will attract 'eyeballs' while giving little or no space to equally or perhaps even more important topics.

Credibility

Generative AI tools such as OpenAI's ChatGPT and DALL-E 2 enable major publishers to generate content at lower cost and in a fraction of the time it would take a human to do so. This is a potential problem once we reach a stage where AI engines are selecting content created by other AI engines, leading in the end to undifferentiated reporting across media platforms, while making it even easier than it is already to manipulate media coverage.

NewsGuard has identified 49 websites across seven languages that are entirely or mostly generated by AI language models designed to mimic typical news websites. Some of the content on these websites represents serious misreporting (Stefan Kelly, 2023). Clearly, the media sector has its own massive challenge: how to distance itself from elements that spread fake news and attempt to manipulate the public.

Avoiding Bad News

More and more people feel overwhelmed and fatigued by the continual flow of bad news, resulting in news avoidance. Women are turning away at faster rates than men, and narrowly focused media channels like The Know are focusing on the female segment in an attempt to help transform their relationship with the media.

But the hunger for good news is a problem for mainstream media. For example, in the first few months after launch the first episode of actor John Krasinski's 'Some Good News' YouTube series was viewed more than 17 million times (Erin Levitsky, 2020). Social media can attract large audiences simply by promising 'good news', which would contribute further to the dumbing down of the very audience that needs to understand the challenges we are facing and how they can contribute to addressing them.

What's in It for the Media?

The news and current affairs media are losing audience share to social media, a worrisome trend given that media organizations rely on advertising and subscriptions for revenue (Hannah Uguru, 2023):

- In 2022, just 35% of Britons who said they regularly use social media for news agreed that these platforms are trustworthy.
- Just 9% of people in the UK pay for online news.
- 17% of 18- to 24-year-olds say they 'mostly watch' (rather than read) the news online, compared to 11% of over-55s.

Again, individuals potentially play an important role, by supporting media organizations that make an effort to ensure their content is free of 'fake news' or simply careless reporting. Ultimately it is the choices of individuals that 'make-or-break' media organizations, as major advertisers will channel at least some of their advertising expenditure to media outlets with attractive subscriber or readership numbers. It follows that individuals like you have an important role to play when it comes to keeping the media honest and ensuring that AI won't take over and feed us potentially unqualified, misleading 'news'.

Non-Governmental Organizations (NGOs)

Not surprisingly, non-governmental organizations (NGOs) are actively contributing to addressing some of the challenges we are collectively facing, often collaborating with companies:

Addressing food security
- The Co-op is making donations to local charity organizations linked to the sales of own-brand products. These donations will be matched by Crowdfunder (Avinash Akhal, 2022).
- Deliveroo and the Trussell Trust launched a scheme to collect in-date food items from customers' homes and deliver them to food banks (Sophie McKay, 2022).
- Earth & Wheat redistribute misshapen or overproduced food to charity partners and consumers via a range of subscription boxes, thus avoiding the generation of food waste (Kathryn Morrisby, 2023).

Challenging government
- UK housing and homelessness charity Shelter has launched a campaign highlighting the unhelpful advice and quotes from politicians such as 'Reuse your old teabags', 'Cancel your Netflix account' and 'Just work more hours' – calling on the government to take meaningful action like making housing more affordable (Makua Adimora, 2022).

Contributing to the Circular Economy
- Charity shops are experiencing a boost in demand as sustainably minded and fast fashion-fatigued young consumers seek an alternative shopping experience (Roisin Lanigan, 2022).

Why do they engage?

- 90% of Deliveroo users want a country where food banks are no longer needed – no wonder Deliveroo joined forces with the Trussell Trust to address food insecurity.
- 9% of parents anticipate needing to visit a food bank to cope with the cost-of-living crisis, with 88% stating that their monthly food bill has increased. Food insecurity is high on their agenda and a party addressing this challenge can expect to get their support.
- 42% of Britons believe that not wasting food is an important part of living more sustainably and reducing their environmental footprint.
- In 2020, 83% of people in the UK said they wanted to get involved with their wider community. Joining an NGO is an effective way of doing this (Sophie McKay, 2022).
- 65% of people would like to be more eco-conscious, but feel unable to do so due to the cost-of-living crisis. But they can support an NGO that is addressing this issue (Kathryn Morrisby, 2023).

It is getting a bit repetitive, but again the point needs to be made that individuals have an important role to play. Their support of NGOs – which can be shown through posting supportive content on social media, volunteering, or donating goods or money – makes a difference to the organization's ability to address major challenges, to find partners with resources to collaborate with, and to gain government support.

Government

Governments have massive resources as well as the power to regulate and legislate. For this reason, they can play an instrumental role when it comes to addressing major global challenges. Here are a couple of examples, showing actions enlightened governments have taken:

Circular Economy

The French government has announced a policy allowing people to claim back up to €25 from the cost of their clothing or shoe repairs as of October 2023. People can reclaim between €6 and €25 of the cost of any repairs that are made in participating workshops and cobblers. €154 million has been set aside to fund the program for five years (Nick Lising-White, 2023).

Media literacy

The Finnish government has established training programs on media literacy in schools. Finland is teaching students how to identify false information and propaganda, starting at pre-school level. The government is also using libraries as centers for teaching older generations how to identify misleading online information (Media Literacy Finland, n.d.).

Why does the government care?

Of course, the 'right' answer is that the government is supposed to care – it's the government's obligation to address major challenges. However, we have already quoted several instances where the government was not guided by a commitment to better the life of its citizens but rather by monetary incentives, values that were not aligned with the electorate, or by simply giving way to lobbyists. It is therefore especially rewarding to see actions such as the Finnish government's media literacy program.

Survey data shows that governments in many countries need to work towards becoming credible, reliable, and transparent leaders that address the challenges their citizens suffer from. It is noteworthy that, globally, businesses hold a 54-percentage point lead over governments in terms of perceived competence, and also that they are 30 points ahead on ethics (Flavia Russo, 2023).

Here, of course, individuals living in democracies have the most direct impact because they have a vote when it comes to which party will earn the right to govern.

The Power of the Individual

While they may not have the resources or funds to help directly address the massive challenges we face today, individuals do have more power than they may realize. They can certainly influence the actions of industry leaders, politicians, media owners, and decision-makers in NGOs by the choices they make when it comes to what they buy and where they buy it from, which organizations and brands they support on social media or when making recommendations, which challenges they encourage others to help address, who they vote for, and, in general, how they use their informal influence.

Importantly, individuals can make it their business to be well-informed and to fight misleading media reports and social media content. We need more media vigilantes who fight for the truth, as only an informed citizenship will be able to take considered action.

All of us can make a difference. For some it will be easy; for others it may be more challenging. But nobody should say that they are too insignificant to make a difference, so they may as well not try. Every action, every vote, every purchase, every statement counts when it comes to shaping the actions of those who have the resources and influence to formulate policy and instigate change on a macro level.

This is the power of the individual. This is the power you have regardless of the resources and power you can bring to bear.

Future Research Using the DMA/DMS When Viewing Complex Intersectional Issues

As well as being a book about how we act in the face of significant global and more local challenges, the book also set out to illustrate a specific methodology for understanding these issues. Our selected methodology was the Declarative Mapping Approach (DMA) and its major tool, the Declarative Mapping Sentence (DMS) (Gordley-Smith and Hackett, 2023; Hackett, 2014, 2018, 2019, 2020, 2021 Hackett and Gordley-Smith, 2023; Hackett and Lustig 2021; Lustig and Hackett, 2020a, 2020b). We chose to employ the Declarative Mapping Approach because the Commitment Gap that we were investigating consisted of multiple powerfully interacting components or aspects. Such components were features of individuals or groups, situations and aspects of the issues with which we were concerned. With such an enmeshed complexity we needed a methodology that would capture the lived, real-life interactive network that was formed by these different variables. Furthermore, the information we were evaluating and presenting was typically qualitative secondary data. The nature of this information severely delimited the choice of methods available to us. However, the Declarative Mapping Approach appeared to fulfill the needs we had and we used this to structure the reviews we performed and the understanding we developed in our writing.

We believe that the writing of declarative mapping sentences for specific challenges that we face was an extremely useful way to explicate the intricacy of the environmental issues we presented. The DMS is able to preserve the complexity of real-world events and issues whilst displaying these in a manner that enables raiders of the sentences to more clearly appreciate the issues being addressed. Such appreciation takes the form of providing a better understanding of the major components of an issue or event in the form of the facets in mappings sentences and the internal make-up of major components is revealed in each facet's elements. The issue or event is clearly presented in the declarative mapping as a whole and the connective phraseology that is used to connect the facets suggests the way in which the facets relate to each other and how they come together in the real-world to form the issue or event of concern.

We claim in this book that by using a Declarative Mapping Approach the researcher guards against over-simplifying an event for the sake of being able to present and investigate the issues with which they are concerned.

As mentioned at the start of this final words section of the book, the authors are at present engaged in work that aims to develop our understanding of social issues and environmental issues. Values are aspects of life that we hold to be important. There can be little disagreement that the issues we have presented in the forgoing pages, along with many other issues we have not mentioned, are of great individual and social importance.

Throughout the book we have emphasized the importance of the just, fair and equitable use and development of the resources on the earth. Consequently, social justice and environmental justice are of prime importance in all of our current and future research.

As a more detailed example, the third author is conducting research regarding the development of a multidimensional psychological framework for understanding the relationship and intersection of attitudes toward environmental and social justice issues (Crenshaw, 1989). She employs Facet Theory and the Declarative Mapping Approach to design and analyse the related qualitative data with the aim of applying this framework to develop more effective future social justice and environmental justice strategies. An example of her working DMS is below:

EXAMPLE 7.1   AVA GORDLEY-SMITH'S INITIAL DMS FOR UNDERSTANDING THE INTERSECTIONALITY OF ENVIRONMENTAL ISSUES AND SOCIAL ISSUES

This study aims to understand the latent psychological frameworks people may have towards issues of ecosocial justice and how those related thoughts, emotions and actions may reflect in having a/an:

Attitudes

positive
neutral
negative
undetermined

attitude, and a/an:

Behaviors

positive
neutral
negative
undetermined

behaviors toward issues of:

Environmental

desertification
land degradation

rising sea levels
deforestation
biodiversity loss

and issues of:

Social

racism
classism
sexism
genderism
heterosexism
ageism
ableism

in order to see if they perceive these environmental and social justice issues to be:

Intersectionality

intersecting with
independent from
dependent on
undetermined relationship with

one another, and through the application of an eco-social intersectional framework these relationships lead to:

Impact

more
less

the same level of understanding.

References

Adimora, Makua, Shelter Mocks UK Government's Unhelpful Saving Tips, Canvas8, September 29, 2022.

Akha, Avinash, Seanack Campaign Gamifies Beach Clean-ups, Signal, Canvas8, August 10, 2022.

Akhal, Avinash, Co-op x Crowdfunder Blends Donations with Daily Habits, Canvas8, November 18, 2022.

Bosky, Nicole, Airbnb Supports Fleeing Ukrainians by Providing Housing, Canvas8, March 8, 2022.

Clifford, Eva, Twig: Encouraging People Into the Circular Economy, Case Study, Canvas8, January 18, 2022a.

Clifford, Eva, Twig: Encouraging People into the Circular Economy, Canvas8, June 26, 2022b.

Costa, MaryLou, How SMEs Are Getting Creative to Help Ukrainian Refugees, *Raconteur*, April 08, 2022. https://www.raconteur.net/global-business/small-business-ukraine-relief-effort

Crenshaw, Kimberle (1989) "Demarginalizing the Intersection of Race and Sex: A Black Feminist Critique of Antidiscrimination Doctrine, Feminist Theory and Antiracist Politics," *University of Chicago Legal Forum*, Vol. 1989, Article 8.

Davey, Bruno, Top Multinationals Offer Refugees in Europe Steady Jobs, Canvas8, June 26, 2023.

Gordley-Smith, A., and Hackett, P. M. W. (2023) African Philosophy-Based Ecology-Centric Decolonised Design Thinking: A Declarative Mapping Sentence Exploration, *Filosofia Theoretica: Journal of African Philosophy, Culture and Religions*, V12(2). https://www.ajol.info/index.php/ft/article/view/260077

Greenblatt, Josh, Levi's: Courting Gen Z with Sustainable Values, Case Study, Canvas8, December 6, 2022a.

Greenblatt, Josh, Levi's: Courting Gen Z with Sustainable Values, Canvas8, June 26, 2022b.

Hackett, P.M.W. (2014) *Facet Theory and the Mapping Sentence: Evolving Philosophy, Use and Application*, Basingstoke: Palgrave Macmillan Publishers.

Hackett, P.M.W. (ed.) (2018) *Mereologies, Ontologies and Facets: The Categorial Structure of Reality*, Lanham, MD: Lexington Books.

Hackett, P.M.W. (2019) *The Complexity of Bird Behaviour: A Facet Theory Approach*, Cham, Switzerland: Springer.

Hackett, P.M.W. (2020) *Declarative Mapping Sentences in Qualitative Research: Theoretical Linguistic, and Applied Usages*, London: Routledge.

Hackett, P.M.W. (2021) *Facet Theory and the Mapping Sentence: Evolving Philosophy, Use and Declarative Applications*, (second, revised and enlarged edition), Basingstoke: Palgrave Macmillan Publishers.

Hackett, P.M.W., and Gordley-Smith, A. (2023) A Declarative Mapping Analysis of the Philosophy of Paulin Hountondji, *Ogirisi International Journal of African Philosophy*. https://www.academia.edu/77109999/An_Exploration_and_Exposition_of_Paulin_Hountondji_s_Philosophy_Through_a_Declarative_Mapping_Sentence_Approach

Hackett, P.M.W., and Lustig, K. (2021) *An Introduction to Using Mapping Sentences*, Basingstoke: Palgrave.

https://es.unesco.org/creativity/policy-monitoring-platform/media-literacy-finland-national (n.d.).

Kelly, Stefan, AI Content Farms Threaten Trust in Online Media, Canvas8, May 10, 2023.

Kenyon, Riani, Tesco Sets Up, Reverse Supermarket Amid Food Crisis, Canvas8, December 1, 2022.

Kenyon, Riani, Murfy Offers Hassle-free Repairs to Home Appliances, Canvas8, October 9, 2023a.

Kenyon, Riani, Mastercard Helps Ukrainian Refugees Resettle in Poland, Canvas8, July 4, 2023b.

Lanigan, Roisin, How are Gen Zers Fuelling the Charity Shop Boom?, Report, Canvas8, May 24, 2022.

Levitsky, Erin What Do People Want from the Media in Anxious Times, Report, Canvas8, April 30, 2020.

Lising-White, Nick, French Government Gives Back to Sustainable Shoppers, Canvas8, July 21, 2023.

Lopez, Nicolas, Selfridges, Swap Shop Makes Circular Fashion Easy, Canvas8, July 21, 2023a.

Lopez, Nicolas, Selfridges, S Swap Shop Makes Circular Fashion Easy, Canvas8, June 26, 2023b.

Lustig, K., and Hackett, P. M. W. (2020a) *The Philosophy of Facet Theory Pocket Guide*, San Francisco: Blurb Publishing.

Lustig, K., and Hackett, P. M. W. (2020b) *Mapping Sentence Pocket Guide*, San Francisco: Blurb Publishing.

McKay, Sophie, Deliveroo x The Trussell Trust: Tackling UK Food Poverty, Case Study, Canvas8, May 26, 2022.

Morrisby, Kathryn, Earth & Wheat: Wonky Food for a Cost of Living Crisis, Case Study, Canvas8, March 2, 2023.

Russo, Flavia, Laura Dodsworth on the Future of Communication, Report, Canvas8, September 29, 2023.

Sambell, Luana, Supportive Action Gives People Faith in Brand Values, Canvas8, March 18, 2022.

Steidl, Peter (2023) *Fearless Brands; Successful Brand Strategies for the Fear Economy*, Hannover, Germany: Kochstrasse Enter Tomorrow Publication Series.

Uguru, Hannah, What's the Future of Independent and Mission-led Media?, Report, Canvas8, June 27, 2023.

Zadeh, Leila, Burger King Potato Giveaway Supports French Farmers, Signal, Canvas8, February 26, 2021.

INDEX

ability 9, 13, 44–45, 61, 65, 68, 99, 114
able 2, 5, 13, 32, 35, 38, 43, 51, 65,
 86–87, 97, 100, 111, 119, 128–129
accept viii, x, 36, 46, 52, 57–58, 60, 69,
 82, 95, 98–99, 101, 113, 124
across x, 25, 44, 68, 71, 77, 88, 100,
 122, 125
action viii–xii, 1–2, 5–6, 8–13, 15, 17,
 19–20, 34–36, 43–47, 51, 57, 59,
 65–66, 69, 72–73, 77, 80, 95, 97–98,
 100, 102–103, 106–107, 109, 112,
 116–131
activities 15, 20, 31, 36, 43, 48, 54, 65,
 71, 95, 100
address viii–xiii, 1–3, 6, 9–10, 13, 22,
 24, 26, 30, 35, 39, 41, 45, 55, 60,
 62, 66, 69, 75, 79, 93, 97, 99–100,
 109, 111, 113, 115
addressing viii, xii, 1, 3, 9, 11–12, 33,
 57–89, 93, 109, 111–112, 120,
 122–123, 125–127
affected 1, 8–9, 19, 22, 24, 33, 36,
 40–41, 51–52, 64, 105, 113
Africa 18, 29, 35, 62, 78
African 29, 52, 59, 62
after xii, 3, 7–9, 13, 25, 27, 29–30,
 32–33, 39, 43–44, 46, 52, 54, 67, 75,
 81–82, 84, 97, 102–103, 109–110
again ix, 15, 17, 19, 30, 32, 39, 44–45,
 48, 53–54, 71, 73, 95, 106, 112–113,
 122, 126–127
against viii, 9, 14, 22, 24, 52, 60, 73, 129

age xi–xii, 1, 14, 30–31, 68, 77, 83, 113
agenda 12, 15, 93, 95–96, 99, 105,
 112–113, 127
agricultural 25–26, 74, 77–78
agriculture 25–26, 35, 52, 77
AI 79–80, 125–126
alcohol ix, 30–31, 35, 38, 48, 68
allow 22, 43, 48, 51, 64–66, 78, 86–87,
 100, 106, 113
allowing 10, 12, 19, 30, 43, 65, 77–78,
 103, 116, 118, 122, 127
allows 43, 50, 65, 69, 77, 98
already viii, x, 16, 19, 45, 48, 64,
 86–87, 97–98, 102, 105–106, 114,
 120, 124–125, 128
alternative 16, 34, 48, 54, 95–97, 126
always 2, 8, 19, 40, 55, 57, 69, 84, 86,
 99, 101, 106–107, 112–113
American 26, 52, 63, 65, 98, 101, 106
Americans 15–16, 41, 49, 63, 65, 101
analyse xiv, 43–44, 67, 72, 78, 88, 109,
 116, 130
another xi, 9, 13–14, 22, 28, 31–32, 41,
 43, 45, 48, 52, 54, 60, 65, 70–71, 81,
 84, 97–98, 100, 103, 105–106, 110,
 122, 131
answer x, 30, 58, 60–61, 83, 85, 97, 99,
 113, 119–120, 128
anyone 2, 13, 30, 34, 52–53, 63, 68, 79,
 84
anything 3, 15–17, 22, 44, 82, 110, 112
apply 32, 49, 61, 65–66, 83, 116

approach ix–x, xii–xiii, 2, 13, 27–28, 30, 33–36, 39, 47, 51, 53–54, 57–58, 61, 69–71, 95–97, 105–106, 109, 116–118, 129–130
April 14, 17
area xii, 24–25, 34, 52, 103
areas ix, 8–9, 14, 25, 33, 54, 64–65, 69, 73, 112
around 10, 19, 26, 29, 31, 35, 39, 46, 48, 57, 61–62, 65, 70, 73, 85, 96, 103, 111
asked 8, 71, 83, 95, 105–107, 119
aspects xiii–xiv, 41, 85, 113, 129
attention 9, 13, 19, 30, 44, 46–47, 82, 95, 99, 110
attitudes 8, 26, 61, 110, 113, 124, 130
Australia 12, 24–26, 33–35, 50, 52, 54, 57–59, 64, 75
Austria 27–28, 51, 53, 70
Austrian 27, 51, 53–54, 70, 86–87, 98
authors xii, 3, 27, 62, 82, 95, 103, 129
available 20, 59, 61, 66, 75, 84, 86, 107, 110, 113–114, 129
average 16, 26, 28, 38, 68, 75, 82, 85, 105, 107
avoid x–xi, 13, 15, 25, 38, 46, 48–50, 57, 66, 79–80, 88, 95, 97, 107, 112
aware 1, 3, 43, 55, 71, 100, 103, 112, 116
awareness 4, 16, 67, 72, 75, 78, 88, 95

back ix, 48, 50, 54, 62, 67, 72, 75, 78, 81, 84–88, 99, 103, 111, 123, 127
bad x, 8, 41, 46, 80, 83–84
balance 9, 17, 99, 106, 110, 114–115
based ix, 8, 14, 41, 48, 53, 58, 62, 71, 75, 80, 82, 97, 101
basic 13, 35, 48, 55, 59, 80, 82, 84–86
become 3, 9, 22, 31, 39, 41, 45, 49, 58, 60–61, 73–74, 88, 96, 103, 113, 124
before 2, 13, 25, 29, 32, 44, 49–50, 71, 73, 107, 112–113, 115–116, 118, 122
behaviour ix–xii, 8, 31, 35–36, 38–39, 43–45, 47–55, 58, 65, 95, 97–98, 102–103, 107, 113–114
behavioural ix, 38–39, 44, 61
being x, xii, 3, 5, 13, 16, 19, 28–29, 33–34, 39, 41–44, 46–47, 50–53, 61, 68, 74–75, 81, 84, 93, 101, 106–107, 111, 116
belief viii, 4, 9, 32, 34, 41, 53, 67, 71–72, 78, 88, 107

believe 3–6, 8–11, 13, 18, 20, 38, 40–52, 58–59, 67, 69, 75, 85, 98–99, 110–111, 115
belonging 34, 36, 50, 54–55, 61, 101–102, 107
benefit xii, 6, 33, 36, 59, 61, 65, 69, 77, 83–84, 87, 116
benefits 6, 14, 19, 26–27, 43, 67, 77, 82–87
between xii, 3, 5, 13, 18, 22, 28, 33, 35, 41, 44, 47, 49, 51, 58, 69, 85, 98, 101, 109, 111, 113–114
billion 33, 35, 58, 62–64, 73–74, 77–78, 84, 109
board 28, 66, 88, 97
book x, xiii–xiv, 4, 13, 40, 100
boost 25, 35, 47, 77, 107
brain ix, 9–10, 43–45, 47–55, 65, 102, 105
bring 40, 50, 54, 57, 66, 82, 97, 99, 107, 115
build 28, 54, 62, 64, 81, 99–101, 111, 115
building 1, 5–6, 26, 34, 39, 67, 101, 116
business 14, 28, 35, 60, 77, 79, 85
buy ix, 8, 30, 45, 58, 103, 105, 113–114

campaign 15–16, 32, 38–39, 103, 114
camps 2, 24–25, 35–36, 51, 53, 58–62, 110–111
care 10, 28, 35, 46, 54, 58–59, 61, 65, 69, 86–87
case x–xi, 20, 24, 27, 32, 36, 40, 50, 52–55, 61, 63, 69, 71, 84, 98, 101–102, 105–107, 113
cases x, 13, 22, 35, 49, 77, 80, 102, 105, 114
catastrophic viii, x–xi, 2, 9, 45, 96, 110
cause xi, 34, 49–50, 100, 106, 112
challenge viii, x, 1–2, 4–5, 9–10, 12–13, 15, 33, 35, 45, 49, 51–53, 57–61, 64, 66–69, 74–75, 77–80, 82, 88–89, 96–97, 99, 106, 109–110, 112, 115–116
chances ix, 40, 42, 44, 47, 49, 61, 93, 114
change viii–ix, xi, 1, 9, 12–19, 24, 29–30, 33–35, 38–39, 43, 45–47, 51, 54, 57–58, 63–67, 74, 81, 93, 95–100, 107, 110–111, 114–115
changed 6, 17, 24–25, 27, 29, 52–53, 98, 106, 113

changes 2, 35, 46, 55, 78, 99, 107, 109–111
children 30–31, 58–59, 61, 66, 68, 73, 83, 110–111
China 6, 18, 24, 28, 49, 64, 77, 96, 98–101
circular 122, 124, 126–127
cities 1, 13, 15, 31, 40, 45, 63, 65, 69–71, 73
citizens xi, 2, 40, 46, 58, 60, 62, 71, 78, 80, 82, 86, 100, 103
city 25, 40, 53, 69–72, 86, 103
claim 2, 6, 28, 80, 96, 98–99, 113
claims 10, 14, 25, 40, 46, 54, 62, 70, 77
climate viii–ix, xi, 1–2, 6, 9, 11–20, 24, 34–35, 45–46, 52, 54, 63–67, 74, 93, 95–99, 110–111
club 31–32, 36, 44, 50, 54, 101
coal 12, 25, 33–36, 54, 78, 102
collective ix–x, xii, 1, 9, 12–13, 40, 57, 111
Commitment Gap ix–xiii, 1–20, 35–36, 54, 57, 71, 100–101, 106, 111–112
community 18, 20, 26, 33, 40, 45–46, 53, 57, 69–71, 83, 101, 103
companies 16, 28, 34, 54, 74, 80, 86, 96, 101, 110
company 28, 77
compared 20, 62, 70, 73, 84–85, 102, 105
complex xii, xiv, 2, 4, 26, 34, 41, 43, 55, 63, 70, 105, 110
concept 15, 31, 44, 66, 70, 87, 101, 115–116
concern 1, 14, 19, 22, 53, 55, 103, 106
concerned 14–15, 17, 41, 44, 52, 58, 96, 113
conditions 19, 27, 35, 58, 61–62, 65, 71, 78, 110
conference viii, 1, 6, 15–17, 46, 52, 54
consequences ix, 15, 28, 38–39, 43–45, 62, 79, 101, 110
consider xii, 2, 10, 34, 36, 41, 49, 86, 99–100, 102, 106–107, 113, 115
consumers ix, 45, 75–77, 88, 105
contribute 19, 77, 97, 100, 120–121
COP 17–18, 46, 54
cortisol 48–50, 52–53, 55, 61, 66, 101–102
cost 58, 63, 69–70, 81, 86–89, 97, 103
countries x–xi, 10, 17–19, 24–25, 27–28, 31, 33, 35, 41, 52, 57–62, 64–65, 67–75, 77–79, 82, 87, 89, 97, 100, 113

COVID 19, 38, 41, 46, 61–63, 68, 109
create 9, 32, 34, 46, 48, 50–51, 63, 78–80, 88, 100–101, 106–107
created 24, 48, 63, 81, 99, 101, 106, 110
creating 27, 50, 65, 87, 93, 97, 99–100, 102, 105, 107, 115
climate crisis viii–xi, 1, 6–8, 11–13, 15, 24, 35, 57, 67–69, 74, 82, 89, 96, 99, 106
critical 6, 28, 34, 44, 67, 72, 79–80, 87–88, 95
customers 105–107
cut 58, 80, 83–87

danger 2, 9, 14–15, 39, 44, 52, 62, 65, 110, 112
data x, 77, 80, 89, 113
deal x, 9, 13, 16, 28, 30, 33–34, 43, 47, 49, 60, 66, 69, 78, 82, 96, 115
dealing 2, 24, 30–31, 33, 36, 45, 60, 64, 82, 89, 112, 114
decades viii, 1, 6, 13, 20, 27, 29, 39, 59, 64, 99
decisions 8–9, 28, 41, 46, 54, 100, 103, 107
decisive ix–xii, 1–2, 5–6, 9–13, 17, 19, 40, 49, 69, 80
declarative mapping approach xii, 129–130
declarative mapping sentence xii–xiii, 4–5, 67, 72, 78, 88
decline xi, 70, 85, 87
deliver ix, 2, 47–48, 52–53, 65, 80, 85, 87, 106
delivered xi, 14, 19, 36, 39, 46, 65, 86–87, 95
demand 9, 20, 25, 64, 95
demonstrations 14, 19, 93–97
denial 5, 11–14, 16, 71, 82
department 68–70, 73, 109
designed x–xi, 6, 9, 19, 25, 44, 53, 62, 66, 100, 102
desire 13, 44, 47, 50, 66, 106, 112, 115
despair 6, 9, 12–13, 19, 111
destroy xii, 9–10, 20, 35, 79–80
detention 24, 35, 51, 58, 60, 89
develop 4, 16, 25, 45, 48, 57, 61–62, 65, 79, 81–82, 99, 103, 112, 115–116
developed xi, 1, 3, 24, 43–44, 49, 52, 55, 57, 61–62, 69–70, 75, 77, 81, 99, 114
developing xi, 34, 43, 45, 57, 61–62, 78, 82, 103, 107, 111

development xi, 9, 16, 22–26, 54, 59, 63–64, 66–70, 72–73, 78–79, 81, 87–89, 101
difference 3, 34, 41, 65, 99–100, 111, 113
different ix–x, 1, 3, 6, 8, 22, 31, 35, 39, 55, 57, 69, 75, 83, 93, 106, 114
difficult 2, 19, 31, 35, 54, 68, 73, 78, 114
directly ix, xiii, 15, 35, 44, 52, 83, 98
displaced 25–26, 35, 59, 63, 85
DMS xii–xiv, 4–5, 11–12, 67, 72, 78, 88
donations ix, 35, 44, 50, 105–106
dopamine 10, 20, 30, 36, 47–55, 61, 66, 102
drink 27, 38–39, 102
drive 10, 39, 44, 54, 95, 99, 105
drivers x, 38, 55, 74, 80, 114–115
drives ix–x, 38–39, 43, 47–49, 55, 98
driving ix–xi, 38–39, 43–44, 48, 55
due 19, 22–24, 29, 33, 36, 40, 43, 52, 54–55, 59, 61–63, 68, 70, 74–85, 97, 110

Earth xi, 2, 12, 14–18, 29
easier 60, 69, 99, 114–115
economic xi, xiv, 6, 12, 17, 20, 28–29, 33, 45, 54, 60, 68, 73, 77, 81, 85–86, 97, 101, 109–110
economy 25, 27–28, 33, 45, 85–87, 101, 113, 117
effect xiii, 15–16, 20, 44, 103–105
effective 4, 15, 30, 36, 39, 45, 48, 51, 53, 65–67, 72, 79, 86, 89, 95–96, 106–107, 115
effectively 1–3, 13, 22, 27, 34, 41, 65–66, 109
effort xii, 8, 10, 35, 44, 59, 71, 81, 99, 106, 109, 111
either 16, 30, 41, 47, 59, 61, 80, 82, 97, 107
electricity 25–26, 33, 35–36, 50, 52, 63, 78, 101
eliminating 57, 75, 79, 88–89, 116
emissions 2, 14–15, 18–19, 34, 55, 63, 97
employees ix, 27–28, 86–88
employment 24, 26, 33, 35–36, 66, 81, 83, 86
energy xii, 2, 8, 19, 26, 33–34, 36, 41, 49, 54, 110
environment ix, 9–10, 16, 20, 35, 43–45, 47–49, 55, 59, 95, 102, 107, 109–110

established 15, 25–26, 29, 33, 35, 40, 49, 52–53, 103
EU 24, 28, 49, 74, 77–78
Europe 14, 19, 25–26, 31, 68, 77–78, 80
European 24, 26, 29, 31, 52, 58–59, 65, 70, 73, 77, 80
events 2, 8, 13–14, 17, 19, 24, 32, 40, 45, 48–50, 55, 64, 70, 100
evidence ix, 1–2, 14–15, 28, 40, 55, 71, 82, 84, 97, 100, 112
evolving xi, 1–2, 8, 13, 43
experience xii–xiv, 9, 47–49, 59, 61, 64–65, 74, 98, 101, 107, 112, 114, 116
experiencing ix, 15, 51, 55, 68, 74
experts 18, 54, 66, 71, 110–113
explore 1, 6, 29, 39, 48, 55, 57, 93, 107, 116

face viii, xi–xii, 2–4, 6, 9–11, 13, 15, 29, 35, 43, 53, 67, 71–72, 78, 88, 96
facet xii–xiii
facets xii–xiv
Facet Theory 130
facilitate xiii, 17, 26, 47, 62, 81, 85, 110
fact 2, 10, 43, 50–51, 54, 58, 61–62, 65, 75, 80, 99, 114
farmers 8, 77–78
favour 54, 85, 99, 101, 105, 110, 114–115
fear x, 36, 48–51, 53, 58, 66, 69, 83, 85, 115
feel ix, xii, 3, 5, 9–10, 15, 20, 32, 43–48, 50–52, 55, 66, 69, 84, 102–103, 106, 111
focus ix–x, xii, xiv, 2, 9, 15, 27, 34, 44–45, 71, 73–75, 79, 83, 99, 112, 116
force 25, 34–35, 57–59, 63, 88, 97–99, 101
forward 6, 9, 17–19, 47–48, 57, 67, 72, 78, 80, 84, 86, 88, 99, 116
France 6, 18, 27–28, 53, 70, 75
future viii, xi–xii, 12–17, 19–20, 28, 30, 33–34, 48, 54, 57, 59, 74, 82, 85–86, 93, 99–101, 110–111

gain ix, 43, 45, 71, 87, 110, 115–117
Gap ix–xiv, 1–20, 35, 57, 71, 109, 111, 113
gender x–xi, 1–2, 6, 9, 11–12, 99, 113
generate 26, 33, 51, 54, 80, 106
Germany 17–18, 26, 28, 33–34, 36, 54, 58–59, 64, 75, 78

global vii–xi, 6, 13–20, 22, 34–36, 39, 45, 57–89, 93, 97, 101, 112–113
government 2, 14, 16, 19, 25–27, 29, 33–34, 36, 39, 41, 52–54, 59, 64, 66, 70–71, 77, 82–85, 97–98, 102–103, 115
growing 14, 25, 49–50, 59, 75, 86–87, 93, 101

Hackett 2–5, 20, 67, 72, 78, 88, 100
happy ix, 2, 14, 28, 48, 51, 80, 82, 100, 113
hardwired 9–10, 13, 43–45, 47, 49, 54, 65, 95, 99, 102
Health 24, 39, 49, 62, 65–66, 68, 70–71, 73, 85–86, 106, 109, 112
help ix–xi, 9–10, 30, 32, 39–40, 49, 65, 68–69, 77, 79, 100–101, 106, 111–115
history 6, 13, 22, 49, 53, 58–59, 74, 93, 95, 98–99, 107
home 26, 32, 69, 74, 86–88
homeless viii, 58, 60, 68–71, 73, 75, 110, 113
homelessness viii, x–xi, 1, 11–13, 57, 68–72, 82, 99, 109
hostile 9–10, 15, 44–45, 47–49, 55, 95, 102
House 15–17, 20, 26, 63, 68
Housing 68–71, 122
human viii–x, xii, 2, 4, 9, 14–15, 20, 27, 35–36, 38–39, 41–45, 47–49, 51, 53, 55, 58, 69, 79, 81, 96–97, 111
humankind 9, 22, 40–43, 45, 95, 102
humans x, 2, 10, 36, 40–44, 48–51, 55, 79, 87, 95
hunger xi, 47, 58, 72–74, 78–79, 111

Iceland 30–31, 36, 48, 54, 102
idea 16, 48, 55, 60–61, 63, 69, 82–84, 86, 103, 114, 116
immigration 24–26, 35–36, 52, 57–58, 65
impact viii, x, 8–9, 15, 17, 19, 24, 39, 47, 51, 57, 59, 63–69, 71, 74–75, 79, 81–82, 84–86, 88, 93, 95, 100, 103, 106, 110–111, 115, 117
including 26, 31, 35, 61, 68, 73–74, 82, 97, 111
Income 1, 69, 73, 77, 82–87
increase 5, 9–11, 26, 73, 77, 88, 106, 109, 114
independent 6, 66–67, 98

individual 3, 51, 62, 68, 72, 79, 89, 97, 100
industry 14–17, 19, 26, 34, 54, 61, 78, 82, 93, 96, 110
influence ix, 32, 38–39, 41, 50, 97, 100, 112
information 3, 8, 43, 47, 60, 96, 99, 106, 112, 114
infrastructure 24, 26, 33–35, 63–67, 81, 111
insecurity x–xi, 6, 57, 73, 75, 77–79, 82
international 6, 29, 36, 65, 78, 101
issue xii, 2–4, 6, 8, 15–18, 20, 24–27, 31, 41, 45, 50, 69, 75, 96, 100, 103, 110

job 58, 80, 82–86, 88, 102, 114
justice ix, 2, 4, 6, 8, 10, 12, 14, 16, 18, 20, 24, 26, 28, 30, 32, 34, 36, 40, 44, 46, 48, 50, 52, 54, 58, 60, 62, 64, 66, 68, 70, 72, 74, 78, 80, 82, 84, 86, 88, 96, 98, 100, 102, 106, 109–110, 112, 114, 116

keep 1, 17, 19, 24, 31–32, 35, 38, 41, 58, 77, 83, 86, 93, 103, 116
killed 35, 40, 49–50, 52–53, 59, 80, 111

labour 25–27, 34–35, 54, 58, 63–64, 66, 80, 83, 88
land 22, 64, 74, 99, 103, 130
Law 28, 32–33, 44, 63–64
lead x, 28, 34, 39, 45, 48–49, 51, 61–62, 64, 66, 71, 101, 106, 115–116
leaders 2, 6, 8, 18, 33–35, 53, 59, 82, 93, 96, 110
leading xii, 5, 18, 35, 46, 59, 82, 85, 109, 117
leads x, 3–4, 40, 47, 60, 67, 72, 78, 88, 103, 113
learn x, 13, 22, 24, 30, 97, 99, 112, 114, 116
least 9, 24, 26, 32–33, 43, 60, 68, 93, 95
left xi, 10, 19–20, 24, 33, 36, 52, 54, 62, 66, 73, 102, 106–107, 113, 115
life ix–xii, 1–2, 8, 14, 24, 26, 30, 32, 39, 41, 47, 49, 53, 58, 60–62, 65–66, 68–69, 71, 78, 86, 96, 99–100, 111, 115
limited 2, 15, 25, 28, 44, 59, 64, 68, 74, 79, 83, 85, 87

lives xi–xii, 2, 14, 16, 24, 32, 39, 41, 58, 63, 71, 74, 80, 103, 111
living xii, 5, 9, 22, 35, 69–70, 84–88, 103
local ix–x, 13, 19, 30–31, 35, 63–64, 67, 69, 71, 75, 78, 103
London 40, 64

made 1, 6, 14–15, 18, 26, 28, 30, 32, 34, 40, 59, 61, 68, 73, 75, 81, 86–87, 95–96, 102–103, 110, 114
mapping xii–xiii, 4–5, 67, 72, 78, 88
massive ix, 16, 24, 33, 39–40, 45, 59–60, 63–66, 78–79, 81, 85–86, 88, 97, 101, 109–110
media viii, xii, 1, 17, 19, 47, 51, 60, 71, 82, 93, 96–97, 99–100, 112
medical 26, 58–59, 61, 68, 70
men 22, 26–27, 39, 53, 68, 73
mind ix–xi, 35, 38–39, 43–44, 55, 97, 99, 102, 112, 116
model x, 2–5, 8, 20, 31, 67, 71–72, 78, 88, 112
money ix, 3, 10, 33, 39, 41, 55, 65, 70, 75, 83–84, 97, 106, 109, 111

national 15, 25–26, 29, 62
nations 6, 8, 19, 26, 35, 55, 59, 62, 74, 77, 96, 98–99, 101, 109, 111
natural ix, 9–10, 14, 17, 31, 36, 43–45, 47–49, 55, 77, 95, 102, 111
nature x, xiii, 2, 35–36, 38–41, 43–45, 47, 49, 51, 53, 55, 62, 64, 70–71, 74, 96, 111
need viii–xii, 1–3, 8–10, 13, 17, 22, 26–27, 31–36, 38, 40, 44–45, 50, 52, 54–55, 57, 59–63, 65–67, 69, 74–75, 78–80, 82–85, 87–88, 93–99, 101–102, 106–107, 110–114, 116
negative 5, 10, 17, 34, 43, 46, 49, 63, 69, 74, 82, 114–116
negotiations 27, 29, 34, 51, 53
neurotransmitter 20, 47–48, 50, 53, 55, 61
news 35, 46, 50–51, 62, 73, 96–97, 99, 109

old 27, 43–44, 47, 53, 55, 86, 102
once x, 1, 3, 10, 13, 36, 52, 65–66, 81, 83–84, 96, 100, 113, 115–116
opportunities xii, 33, 44, 47, 50, 59, 64–66, 83, 112, 114
opportunity xii, 30, 44, 59, 83, 102

option 53, 71, 85, 96–97, 105
order xiii, 3, 9, 41, 44, 53, 75
organ ix, 31–32, 35, 38–39, 44, 50, 101, 113
Organization 6, 19, 62, 73, 86, 100, 106
organizations 31, 33, 40, 46, 51, 58, 60, 62, 67, 79–80, 87, 99, 101
outcome xi, 1, 9, 17, 27, 99–100, 114–116

pandemic 19, 41, 46, 50, 62, 68, 80, 109–111
past 30, 41, 48, 52, 69, 95
pay 26, 39, 61, 68–69, 83–84, 86–88, 105
people viii–ix, xi, 3, 5, 8–10, 13–16, 18–20, 22, 24–27, 29–35, 38–44, 46, 48–55, 57–65, 67–71, 73–80, 82–87, 93–95, 97–99, 102–107, 110–116
percentage 30–31, 39, 75, 85–86, 105
personal ix, xi, 2, 9, 13, 34, 36, 40–41, 44–45, 50, 52–54, 62, 87, 100, 110, 112
place viii–ix, xii–xiii, 1, 6, 9, 16–17, 19, 27, 29, 40, 43, 48, 51, 53, 60–62, 70, 81, 85, 101, 105, 107
plan 17, 25, 39, 43, 70, 86, 113, 116
play 6, 8, 98
policies 1–2, 8, 10, 24–25, 46, 60, 77, 100
policy 18, 26, 28, 35, 52, 57–59, 69, 85
political viii–ix, 2, 8, 10, 15, 29, 35, 41, 45–46, 52, 58, 60, 78, 99, 109, 112–113
politicians 10, 16, 20, 25, 34, 41, 46, 58–60, 62, 71, 79, 82, 93–97, 100–101
population 14–15, 25, 34, 40, 59, 62, 69, 73–74, 80, 85–86, 97, 101, 110
power 6, 8, 18, 25–26, 29, 32–34, 36, 49, 53–54, 60, 78, 95, 98, 100–103, 111
problem viii–xi, 1, 9, 13, 15, 20, 24, 30, 43, 45, 47, 51–52, 57, 60–64, 68–69, 71, 73–75, 78, 81, 84, 87, 96, 109–110, 113
problems ix, xi, 1–2, 6, 10, 13, 20, 22, 31, 41, 48, 71, 81–82, 109–110, 114
process xiv, 2–4, 24, 34, 43, 45, 47, 50, 82–83, 98, 106, 115
productivity xiii, 78, 80, 82, 84, 87
programme 30–31, 33–34, 48, 53, 63–67, 70, 83, 86–87, 99, 101

progress x–xi, 1, 3, 19–20, 22, 34, 46, 64, 73, 80, 93, 95–97, 99–101, 103, 105, 107, 110
progressive 87–88, 98, 113
provide ix, 26, 33, 54, 60, 75, 82, 112
providing xi–xii, 26, 50, 69–70, 77, 85, 112
public 14–16, 18, 22, 28, 38–40, 46, 50, 60, 63, 65–67, 70, 77, 82–83, 96–97, 103

quality x–xii, 1, 45, 66, 71, 79, 86
question 22, 41, 47, 60, 75, 85, 100, 112, 114, 116

rate 9, 24, 45, 49, 51, 62, 79–80, 83–84, 106
reality 9, 30, 39, 80, 82, 93, 99, 113
reason 3, 8, 14–16, 46, 53, 55, 98, 103, 106, 111, 114
reduce x–xi, 18, 39, 51, 62–63, 66, 77, 84, 111
reducing 2, 19, 38, 52–53, 70, 73, 75, 78, 88, 102, 106
refugee ix–xi, 1–2, 4, 6, 11–13, 24, 33, 57–61, 66–67, 69, 99
relationship 49, 53, 68, 101
report 15–17, 27, 58, 62, 70, 73–75, 80, 84–85
republic 29, 53, 70, 98
research ix, xii, xiv, 1, 3, 15–16, 30, 38, 52, 61, 70, 80, 85, 100, 114
resources xi–xii, 15, 57, 61, 73–75, 100, 111
respect 4, 10–12, 19, 27, 35, 39, 46, 81
respondents 30, 39, 105–106, 113
result 2, 15, 18, 26, 31, 51, 59, 71, 80, 86, 116
results x, 17, 36, 39, 52, 86, 96, 100–101, 105, 107, 113, 116
revolution x–xi, 45, 53, 80–82, 85–87, 89
right ix, 6, 13, 36, 38–41, 44, 47, 51, 60, 69, 81, 93–95, 97–98, 100–101, 114–115
risk xii, 20, 26, 52–53, 60–61, 66, 68, 97
Russia 6, 25, 49, 51–53, 58, 74–75, 97, 101, 111

safe 16, 45, 48–49, 53, 62–63
safety 24, 38, 41, 45, 51–53, 59, 63, 85, 95, 111

save 20, 32, 38–39, 65, 80, 83, 86, 105, 113
scheme 26, 35, 50, 52, 57, 65, 70, 83, 85
science ix, xii, 14–17, 74
service 26, 38–40, 68, 71, 74, 80, 105–106
services xi, 8, 39, 45, 59, 62, 65–66, 69–71, 82–83, 85, 87–89, 96, 105–106
shape x, 1, 28, 36, 39, 43, 48, 52–53, 55, 93, 99, 102–103
shaping 47, 53–55, 101, 115
solution ix, 10, 20, 22, 24–25, 30, 32, 35, 44, 53, 65–66, 69–71, 75, 80, 82, 86, 109–110, 112
solutions xi, 10, 20, 35–36, 52, 55, 57, 60, 71–72, 79–81, 99, 110, 116
sport 31–32, 36, 44, 47
stages 2–3, 5, 12, 29, 100
statistics 24, 68, 73, 81, 111
Steidl 9, 89
step x–xi, 19, 26, 53, 57, 93
stories ix–x, 13, 22, 27, 35, 38, 51, 58, 66
strategies 13, 24, 57, 72, 82
strategy 15, 31, 54, 57, 75, 86, 98, 101, 115
stress 9, 30, 38, 46, 48–49, 51, 61, 66, 68, 102
strong 15, 19, 50–51, 53–54, 96, 100
success ix–x, 4, 13, 22, 30, 33, 35, 38, 51–52, 66, 68, 72, 79, 89, 93, 103, 115–116
successful ix, 14, 35, 39, 55, 59–60, 71, 114
suffer viii, 34, 61, 73, 79, 83, 101
support 3, 15, 34–36, 40–41, 44, 50, 54–55, 58–59, 61, 66–67, 69–71, 77–78, 82–83, 93, 97, 100–101, 103, 106, 109, 114
survival 9–10, 40, 44, 47, 49, 61, 95, 102
sustainable 16, 67, 72, 77, 79, 88
system 6, 9, 17, 27, 39, 41, 44–45, 47, 52, 64, 70, 77, 83, 101, 107, 109

Taiwan 98–99
tax 82, 84–85
technological xi, 14, 20, 45, 80–82, 85–87, 89
threat 2, 9–10, 14, 20, 36, 40, 45, 49–50, 53, 64, 95, 101

today ix–xii, 10–11, 13, 25–28, 30–31, 43, 45–46, 49–50, 53, 58–60, 63, 74, 82, 87, 101, 103
total 36, 59, 75, 77, 116
trigger x, 10, 39, 51, 77, 102, 105, 107
trying xi, xiii, 39–40, 54, 59, 97, 111–112, 114, 116
typically 2, 5, 34, 39, 48, 50, 60, 71, 77, 86, 89, 95–96, 99, 102, 115

UK 16, 27–28, 46, 59, 61, 64, 70, 75, 97
Ukraine 11–12, 24–25, 36, 49, 52, 59, 74, 97, 111
UN 6, 15–17, 19, 58, 73
understand x–xiii, 1–3, 43, 47, 53, 83–84, 98–99, 111, 114, 116
understanding viii–ix, xi–xii, xiv, 27, 43, 48, 98, 115
unemployed 82–85
unemployment 24, 68, 79, 81, 83–85
United Nations 6, 59, 98–99, 111
United Kingdom 6, 26, 53, 59, 68
United States 6, 8, 14–15, 22–23, 41, 44, 49–50, 53–54, 58, 60, 64–65, 67, 69, 73, 75, 77–78, 96–98
universal 29, 82, 84–85
university 70, 85, 89
unlikely 1, 9, 50–51, 64–66, 71, 111
US viii–xii, 2–3, 6, 8–10, 13–20, 24, 27–28, 30–31, 35–44, 46–53, 55, 61, 63–64, 68–70, 75, 77–81, 83, 86, 95–96, 98–101, 103, 109–115

vaccines 8, 61–62
value x, 12, 26, 41, 55, 77, 89, 99, 101–102
values viii, 34–35, 41, 50, 61, 110
views 3, 8, 18, 50, 97, 99–100, 112–113
violence 9, 29, 38, 68–69, 109, 115

war 6, 11–12, 14, 19, 24–25, 29, 36, 40, 49–50, 53, 58, 60–61, 74, 97–98, 101, 110–111
warming 13–20
waste viii, 16, 36, 40, 74–77
water 6, 13, 26, 45, 54, 59, 61, 63, 74, 110
wellbeing xii, 6, 9, 20, 30, 50–51, 97, 107
white 15–17, 20, 26, 29, 36, 80
women 9, 26, 39, 58, 68, 73
words x, 14–15, 17, 27, 36, 48, 54, 69, 80, 84–85, 105, 116
work viii–ix, xii, 2, 13, 15, 25–27, 30, 32, 34, 36, 39, 43–44, 47, 53–54, 57, 65, 79, 81, 83–84, 86–88, 102, 105, 107, 109, 112, 115–116
workers 26–28, 33–34, 54, 64, 66, 79, 85, 87–88, 110
world ix–xii, xiv, 1, 6, 8–9, 13–17, 22, 25, 28–29, 35, 40–41, 43, 46–50, 52–53, 57–62, 65, 68, 71, 73–74, 86, 95, 97–98, 100–101, 103, 109–110

youth 24, 30–31, 48